Mathematical Models for Cell Rearrangement

G. D. Mostow
Editor

Mathematical Models
for Cell Rearrangement

NEW HAVEN AND LONDON
YALE UNIVERSITY PRESS
1975

Library of Congress catalog card number: 74–29731
International standard book number: 0-300-01598-4

Designed by Sally Sullivan
and set in Times Roman type.
Printed in the United States of America by
The Murray Printing Co., Forge Village, Mass.,
Published in Great Britain, Europe, and Africa by
Yale University Press, Ltd., London.
Distributed in Latin America by Kaiman & Polon,
Inc., New York City; in Australasia and Southeast
Asia by John Wiley & Sons Australasia Pty. Ltd.,
Sydney; in India by UBS Publishers' Distributors Pvt.,
Ltd., Delhi; in Japan by John Weatherhill, Inc., Tokyo.

Contents

Contributors

Peter Antonelli *Department of Mathematics, University of Alberta, Edmonton, Alberta, Canada*

Richard D. Campbell *School of Biological Sciences, University of California, Irvine, California*

Narendra S. Goel *Xerox Corporation, Joseph C. Wilson Center of Technology, Rochester, New York*

Richard Gordon *Mathematical Research Branch, National Institute of Arthritis, Metabolism, and Digestive Diseases, Bethesda, Maryland*

M. Lathrop *Department of Mathematics, University of Alberta, Edmonton, Canada*

Ardean G. Leith *Department of Physics and Astronomy, University of Rochester, Rochester, New York*

A. M. Leontovich *Laboratory of Mathematical Methods in Biology, Moscow State University, Moscow, USSR*

D. I. McLaren *Department of Biology, University of Sussex, England*

Hugo Martinez *Department of Biochemistry and Biophysics, University of California Medical Center, San Francisco, California*

I. I. Pyatetskii–Shapiro *formerly of the Laboratory of Mathematical Methods in Biology, Moscow State University, Moscow, USSR, currently unemployed*

Y. B. Radvogin *Laboratory of Mathematical Methods in Biology, Moscow State University, Moscow, USSR*

T. D. Rogers *Department of Mathematics, University of Alberta, Edmonton, Canada*

Robert Rosen *Center for Theoretical Biology, State University of New York at Buffalo, Amherst, New York*

O. N. Stavskaya *Laboratory of Mathematical Methods in Biology, Moscow State University, Moscow, USSR*

Malcolm S. Steinberg *Department of Biology, Princeton University, Princeton, New Jersey*

A. V. Vasiliev *Department of Mathematics, Moscow State University, Moscow, USSR*

E. B. Vul *The Institute of Applied Mathematics of the Academy of Science of USSR, Moscow, USSR*

M. A. Willard *Department of Mathematics, University of Alberta, Edmonton, Canada*

Lawrence L. Wiseman   *Department of Biology, Princeton University, Princeton, New Jersey*

Martynas Yčas   *Department of Microbiology, Upstate Medical Center, Syracuse, New York*

Editor's Note

This monograph presents in a single volume the researches of I. I. Pyatetskii-Shapiro and his collaborators at the Moscow State University Laboratory of Mathematical Methods in Biology, together with related research on cell sorting performed in the West. The first five articles were translated from the Russian by Professor Martynas Yčas. Taken together with the other seven articles and the annotated bibliography, they provide an up-to-date status report on attempts to provide mathematical models for kinetic and morphogenetic features of cell sorting. In selecting the articles for this monograph, I have received valuable advice from my colleague Professor John P. Trinkaus.

G. D. Mostow

New Haven, Connecticut
1974

Introduction

In the progression of the biological sciences, models and facts share certain limitations. Both owe their origin to the whims of an investigator. They result from his perceptions and demonstrate his shortsightedness. Both dwindle in usefulness because of supersedure, and during their more or less brief tenure, they embody a significant amount of the current science.

Yet the relationship between models and facts, and their status vis à vis other aspects of knowledge, is complex. An informal model is a concept that interrelates facts. A formal model is one that has passed somewhat more scrutiny and is also embodied in some kind of expressible language. Thus, models owe their origin and existence to facts. Successful models have enormous predictive value and lead to the renewed collection of new information. Models are, therefore, autocatalytic. It is probably optimal, therefore, for modeling to proceed concomitantly with other theoretical and experimental research.

This volume is an exposition of the use of modeling in examining cell movement, with emphasis on the relations between cell migration and morphogenesis. It is an exhibit rather than a treatise. Twelve articles, which represent two lines of modeling applicable to two overlapping areas of tissue patterning, are reprinted. Although thus restricted in breadth, these selections illustrate the most important properties of a successful model's evolution.

Papers 6–12, for example, represent a continuous train of collective thought photographed at seven intervals in time. In paper 6, the concepts of Steinberg's provocative and exciting differential adhesion hypothesis were modeled and simulated. This attempt suggests that the behavioral properties of the cells may be more critical than had previously been thought. Subsequent papers thus reflect cell motility rules as focal points of attention. Later versions of the modeling (papers 7–9) also escape some of the technical straitjackets of the initial work (e.g., isotropic cells become replaced by anisotropic cells). By the time paper 11 was published, the modeling process had evolved several branches of inquiry, and in this fairly recent paper one finds some of the most original ideas that have developed in the field of cell reaggregation. Concomitantly one observes the fluidity of authors and the changing modeling techniques. Similarly, papers 1–3 show the evolution of an abstract model of cell positioning into studies of morphogenesis in *Volvox*. Papers 4–5 represent models of self sorting similar to the models of papers 7–9, developed independently and simultaneously.

1

One senses that the considerable lack of appreciation of mathematical modeling by many biologists is the fault of the modelers. In too many cases a model is constructed to represent a biological process and a paper is published to describe it; then it is abandoned. The model, at first necessarily abstracted and simplified relative to its biological counterpart, must ripen and develop. After a workable mathematical skeleton is constructed, it must be repeatedly evolved to more nearly simulate the biological phenomenon. Evolution is probably the best sign of a successful and useful model. In this respect the papers in this volume provide a fruitful example because, in two or three groups, they clearly show development of trains of thought and mature from highly abstract to more experimentally applied situations.

A significant theme that runs through all the papers in this volume is the question of whether morphogenetic determinants are global or local. Must one assume that an organism or tissue has ways of assessing its overall shape and pattern and adjusting the behavior of its component cells accordingly? Or can complex patterns arise because cells behave according to personal rules and local configuration? All the papers presented here, and some other recent papers in the modeling literature (see annotated bibliography), explore the assumption of strictly local rules. It seems natural to suppose that, to the extent that cells are the units of morphogenetic pattern formation, they will behave according to their environmental configuration and that this environment is dominated by the surfaces of a very few neighboring cells.

The idea of local rules is relatively new, however, and at least reinforced if not promoted by recent successful modeling. For instance, the early hypotheses concerning the mechanisms of cell sorting concentrated on global aspects of cell arrangement. With Steinberg's differential adhesion hypothesis, one begins to see the decomposition of global and local features of patterning: whereas the overall patterning that occurs in cell reaggregation can be described in global terms, the forces and mechanisms responsible for achieving the overall pattern arise from local operations. Steinberg did not specify exactly what these local operations were, beyond asserting that they involved random cell movements and thermodynamic trapping. Subsequent modeling work, exemplified by this volume, have pursued the problem of local operations both theoretically and by use of simulation. In paper 11, of which Steinberg is an author, we can see considerable advancement in our concept of cell movements and the influential nature of local conditions.

In addition to illustrating how models evolve, this volume provides interesting examples of how independent trains of thought can overlap and use different emphases in approaching similar problems. Finally, to broaden the scope of the book, an annotated bibliography dealing with many areas of the modeling of cell movement is included. Should the reader wish to study seriously the subject represented, I would select the following sixteen articles (listed in the bibliography) as the most useful and broadest introduction to the concepts and methods being explored today in modeling cell locomotion:

Blumenson, 1970 (model of cancer cell spreading)
Blumenson, Bross, and Slack, 1971 (application of cancer metastasis model)
Brokaw, 1972 (computer simulation of flagellar movement)
Carter, 1967 (haptotaxis: passive cell migration)
Cohen and Robertson, 1971 (wave propagation in slime mold aggregation)
Dahlquist, Lovely, and Koshland, 1972 (bacterial chemotaxis)
Gordon and Drum, 1970 (diatom locomotion by capillarity)
Gordon et al., 1972 (rheology of cell sorting out)
Gray and Hancock, 1955 (hydrodynamics of sperm propulsion)
Keller and Segel, 1971 (chemotaxis)
Leontovich, Pyatetskii-Shapiro, and Stavskaya, 1970 (self-aligning of cells)
Lewis, 1973 (cell mixing during development)
Lubliner and Blum, 1973 (non-sinusoidal flagellar waveforms)
Roberts, 1970 (influence of electric and magnetic fields on *Paramecium*)
Silvester and Holwill, 1972 (flagellar waveforms)
Steinberg, 1963 (sorting out of aggregating cells)

Mathematical models in biology enjoy restricted audiences. It is mainly the model builders themselves who appreciate the intricacies and simplicities of a fine model. Biologists generally are not attuned to the current status of mathematical work; they sometimes dislike, distrust, and avoid the published material. One easily definable deterrent to most biologists who value the theoretical work described here is the specialized mathematical notations often used in modeling papers. This volume certainly will not overcome this obstacle, for the papers here are at times recklessly strewn with mathematical notations. Although precise statements are economical to mathematicians who are familiar with them, notation is essentially a formalized jargon and one that is outside the habits even of many biologists well acquainted with the concepts involved. Some readers may deem some of the described models too inadequately related to observed biological processes of cell movement. The ideas of linearization and circularization, for example, may leave many cell watchers blank; on the other hand, these models have basic interest because they represent a natural mathematical first step.

Some maintain that a model is a statement of a new idea, that it represents an extreme probe beyond the current frontier of investigation. When given in its most concise and proper form, a model is not subject to popularization—that is, its formulation in a language common to the average state of the discipline. I suspect that this reasoning has salved the consciences of more than one writer of theoretical biology who has neither the desire nor the energy to make his work more widely palatable. There can be no doubt that the model paper in theoretical biology is that of Turing (1952).* His concept of how strict patterns can arise in homogeneous tissues is innovative, complex, and historically important. The thoughts are strong, yet the paper is comprehensible to anyone. A paper presenting a mathematical model should be clear in statement of concepts and in the relation of the concepts to the biological world. Turing has shown that no model is so esoteric that it can ignore these guidelines.

It is instructive to see which areas of cell motility have been treated mathematically and which areas have not. Flagellar propulsion provides a paradigm for the way in which modeling has aided our understanding of a biological process. For example, it was through the use of modeling that Gray disproved the possibility that flagella act by wagging from the base; Gray and Hancock arrived at a basic understanding of the dynamics of this motion. Then the feasibility of purely local independent bending processes was ruled out on energetic grounds (e.g., by Brokaw); Machin further showed that passive bending of flagellar elements must be taken into account in interpreting waveforms and energy balances. One senses now that most flagellar modeling will shortly achieve a base on data obtained from biochemical analyses of isolated flagellar components. Mathematical modeling and experimental observation have been extremely closely coupled in the study of flagellar locomotion, and in the papers cited in the bibliography of this book, one virtually sees the historical development of our knowledge of this type of cell movement.

Perhaps this situation is not surprising, because flagellar locomotion can be described by relatively few parameters and thus is more biophysical than biological if by the latter term one refers to areas of imponderable complexity. In the most uncharitable view, one might view modeling as a trapping device that isolates biophysical events from our zoo of biological processes. According to this view, flagellar locomotion, particularly its hydrodynamic aspects, is an obvious candidate for immobilization and withdrawal from the undefined.

Other features of cell motility have also been considered mathematically in a profitable way. The subject of cell rearrangements during tissue formation—treated by several papers in this volume—has been significantly aided by mathematical modeling, although a considerable gap remains between experimental observation and theory. For example, it appears that cellular aggregation in the cellular slime mold is becoming greatly clarified by current modeling.

It is surprising to note some aspects of cellular locomotion that have not been modeled. Cellular chemotactic movement has until the last few years not been treated mathematically; this is indeed unexpected because *a priori*

---

* Turing, A. M. (1952). The chemical basis of morphogenesis. *Phil. Trans. Roy. Soc. London* *B*237: 37–72.

chemotaxis appears to be more appropriate for mathematical reconstruction than flagellar locomotion. One problem has been the lack of experimental data concerning the statistical or individual behavior of cells in defined chemical gradients. One should expect, therefore, that the successful laboratory approaches to such data collection and the several attempts to model chemotactic movement (see annotated bibliography) will lead to an explosive investigation into the mathematics of this primordial cell behavior.

A more lamentable gap in the literature of modeling concerns cellular movements during morphogenesis. In the papers collected here and in the literature cited in the bibliography, it is evident that a number of diverse beginnings exist. The problem is obviously one of complexity; morphogenetic movements define three-dimensional, asymmetrical trajectories that are not easily described by manageable expressions. However, the recent work of Gordon and Jacobson on neurulation movements suggests that if one considers the proper cell field and coordinates, simple movement patterns may underlie some complex tissue-shape changes.

A third area of cell movement that has been woefully ignored by mathematical modelers is that of the mechanisms of amoeboid and creeping migration. As in the case of morphogenesis, the migration mechanisms have been treated qualitatively and are currently one of the most exciting fields in experimental cell and developmental biology. As the bibliography indicates, only sporadic attempts have been made to model the nature of cytoplasmic structure or forces responsible for cell shape and shape changes.

Finally there has been scarcely any correlation between models of cell movements and those of pattern formation during embryogenesis. This situation is particularly surprising because the subject of pattern is currently being assaulted by mathematical model builders in grand style. I believe that this irony indicates how little thought has been given to the determinants of cellular orientation and migration that accompany and cause embryogenesis. The developmental biologist's embarrassment at discovering that he has not sufficiently considered how cell movements are patterned during morphogenesis—just as we discover that essentially no kinetic data have been gathered until recently on cell chemotactic migrations—will certainly lead to more experimental attention and modeling along the lines of Carter.

It is regrettable that a volume of this sort, whose subject is cell movement, must bypass the great pioneering works on morphogenetic movements. Unfortunately, these researches have not been sufficiently formalized mathematically to have them fall into this book's domain. But these studies have contributed most generously to our understanding of the motility of cells. One thinks of the fine analytical work of Gustafson and Wolpert on sea urchin gastrulation, Holtfreter on amphibian embryonic cell behavior, Weiss and Garber on cell shape determination, Weiss on contact guidance and other patterns of cell propulsion dynamics, Abercrombie on the ruffled membrane, Ambrose on contact inhibition, and Trinkaus on embryonic cell movement in fish. The fact that these analyses have not been mathematically formalized is in great part a reflection of the multidimensional complexity of cell motility. It is simply not obvious which parameters (if any) of movement can be isolated and used as a framework for a simplified preliminary model. The aspects of cell movement that have been subjected to modeling, exemplified by the papers in this volume and cited in the bibliography, deal with either complex systems where the investigator is interested in only one or a few parameters or with intrinsically simple cases of motility.

On the other hand, the pursuit of global patterns of cellular motility would seem to be both profitable and ripe. The last two decades have spawned several great and perceptive models of biological pattern determination—for example, those of Turing, Wolpert, Crick, Goodwin and Cohen, and Thom. These global patterns are only now beginning to be applied to cell movements, for example by Cohen and Robertson in connection with cellular slime mold aggregation. One may anticipate that significant advances will be made in the near future in modeling cell movement, ranging from the subcellular basis of locomotion to the patterning of movement in populations of cells.

Richard D. Campbell

1. Certain Mathematical Problems Related to Morphogenesis

A. M. Leontovich, I. I. Pyatetskii-Shapiro,
and O. N. Stavskaya*

In examining the geometrical aspect of morphogenetic movements, we are not concerned with the physical-chemical nature of the forces that are responsible for these movements. In our models the motility rules of cells depend only on the position of a small number of their neighbors. The movement in a plane of a line composed of cells is studied. The simplest problem was selected—the problem of straightening a line. This paper presents results of modeling on a computer, which shows that our rules give a solution to the problem of straightening a line.

GENERAL STATEMENT OF THE PROBLEM OF MORPHOGENESIS

The very complex process through which a single cell transforms itself into a multicellular organism has commanded the attention not only of biologists but also of mathematicians. One of the first was von Neumann, who studied the possibility of self-replicating automata.[1] Problems posed by certain effects that occur during embryonic development continued to interest mathematicians. The noted French topologist Thom made an attempt to relate certain problems of embryogenesis to the theory of stratification.[2] Recently, Levenshtein obtained some interesting results on the so-called problems of synchronization.[3,4]

The present paper is devoted to some of the simplest mathematical problems that arise when one attempts to model mathematically the movement of embryonic tissues. Such movements are a significant part of the process of morphogenesis. They are fairly complex, even though only a relatively small number of cells participate (on the order of several hundred or thousand). One of the most studied is gastrulation, which is the pulling inward of parts of a spherical mass (fig. 1). Another movement is the appearance of

Fig. 1. Schematic of gastrulation in the sea urchin.[12]

* First published in *Avtomatika i Telemekhanika* (1970) 4: 94–107.

the neural tube (the primordium of the central nervous system), which is related to the folding of a plane into a tube (fig. 2). There are also movements that form flat surfaces, produce segmentation, and effect other changes.

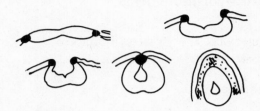

Fig. 2. Development of the neural crest (shown in black) and of the neural tube [after Balinsky, from ref. 7].

One of the approaches to an explanation of such movements is the well-known theory of morphogenetic fields. Although the movements themselves have been fairly fully studied, until now very little is known about the nature of the forces that cause these movements. A very interesting comment on the nature of the morphogenetic movements have been made by Waddington, a prominent embryologist: "[i]t is certain that the elongation of the tissue is not accompanied by an equivalent elongation of the individual cells. It is as though the cells were swept along in the stream of elongating mesoderm and neural tissue like corks floating on a river. The causative agent of the movements must apparently be something at once very small-scaled and transgressive of cell boundaries. Perhaps one could imagine some sort of intimate structure of the protoplasm, such as a fibrisation which was not stopped by the cell membranes."[6] These words were written in 1940, but to this day the situation has been clarified little.

The aim of this work is to study the geometrical aspect of morphogenetic movements and to avoid discussion of the physical-chemical nature of the forces responsible for these movements. In many cases it has been shown that small pieces of isolated tissues are capable of movements analogous to the ones that they undergo as part of the intact embryo.[7,8] It has also been demonstrated that it is not possible to explain the movements of tissues by changes in the shape of cells alone. This leads to the following hypothesis: morphogenesis occurs because each cell has the ability to move in a given direction, which depends on the position of a small number of its neighbors. The motility rules for all cells of a given group must be the same. A consequence of such assumptions about the properties of individual cells is that a portion of the cell mass attempts to achieve the same geometrical form as does the entirety of cells.

Our hypothesis does not exclude the influence of genetic information and of other factors, induction in particular, on the process of morphogenesis.

On the contrary, we conjecture that these factors exert their influence on morphogenesis by changing the motility rules. A test of this hypothesis requires a comparison of biological experiment and mathematical deduction. It is necessary to confirm experimentally that cells do in fact receive information on the geometrical position of neighbors and that on the basis of this information change their rate and direction of movement.

It is worth noting that many types of cells directly inform their neighbors as to similarity or nonsimilarity with themselves. In this respect it is frequently the case that if the neighboring cells are similar, they adhere and form a single tissue; if dissimilar, they disperse. Such mutual testing occurs by means of a special type of sensory proteins located on the cell surface. These proteins have only a limited life-span, of the order of several minutes, but are continually being made under the influence of informational RNA. In this way the recognition system is under immediate genetic control.[9,10] It is very probable that similar systems exist for the determination of the geometrical position of neighbors.

We should also remark that for the time being we do not consider one very important factor—the influence of the extracellular medium—because at the present time very little is known about the properties of the extracellular medium. In principle our hypothesis does not exclude a significant role for the extracellular substrate in the process of movement.

The mathematical verification of the hypothesis formulated above requires concrete construction of rules of cell motility that lead to the formation of tissues of a given geometrical form. This problem leads to the study of certain dynamic systems in multidimensional space. The present work is only an introduction to such a study.

The tissues of interest for embryology are located in three-dimensional space, but because of considerations of mathematical simplicity in the present paper we limit ourselves to one-dimensional tissues (i.e., to lines) embedded in a plane. The discussion below concerns the simplest problem of this nature—the problem of straightening a line. In the following paper another problem will be considered, that of circularizing lines.* We hope that by analogous methods it will be possible in principle to solve problems of constructing in three dimensions such two-dimensional tissues as a sphere or a part of it, a portion of a plane, a cylindrical tube, or a cone. Clearly, it is also possible to construct volumes bounded by the above-mentioned

---

* The problem of circularization was posed by observations of certain events that occur during early embryogenesis. If a portion of a blastula is excised, it will round up into a sphere. Such rounding up into a sphere occurs in many embryonic tissues, particularly those of the skin of amphibians. From considerations of mathematical simplicity we have for the moment decided to confine ourselves to the two-dimensional case; thus arose the problem of circularization. The authors are grateful to L. V. Belousov, who performed the corresponding experiments on the achievement of a spherical form in the laboratory of the Chair of Embryology of the faculty of biology of the State University of Moscow.

surfaces. These mathematical questions will be examined in more detail in later publications.

Nevertheless, the simplest problems discussed below (those of straightening a line and circularization) already raise significant mathematical difficulties. For example, with the help of computer experiments we have shown that certain types of motility rules achieve line straightening and circularization, and we have observed the final configurations, which correspond to a variety of initial states. Nonetheless, a rigorous mathematical proof of convergence toward the final configuration for a sufficiently broad class of rules does not appear to be a simple matter. Also complex is the painstaking demonstration that the empirical rules discovered by us predict such a convergence from the initial states. To a certain extent the matter is related to the following discussion. The movement of a broken line with n apices is described by a special type of nonlinear dynamic system in $2n$ dimensional space, and the investigation of each type of nonlinear system requires its own special methods. To date we have not found adequately effective mathematical methods for the investigation of the special type of dynamical systems that arise in morphogenesis.

Finally, let us note that our approach to the mathematical modeling of embryogenesis differs from the approach used by von Neumann[1] and Thom.[2] In our method the major point is the search for simple rules of interaction between individual cells and their neighbors that make it possible for the entire cell mass to develop in the required direction.

Our approach to morphogenesis was influenced by the work of Tsetlin on the game theory of automata.[11] This theory studies collectivities that solve complex goal-seeking problems yet are composed of separate, relatively simple parts, each of which solves simple problems. Tsetlin arrived at the game theory of automata from considerations involved in modeling the work of neural centers. In the present work we have attempted to apply analogous ideas to the interaction of cells during embryonic movements. We are hopeful that these ideas of Tsetlin's are also applicable to other biological systems.

STRAIGHTENING LINES: RESULTS OF COMPUTER MODELING

We examine a curve lying in a plane and consisting of n segments of straight lines (henceforth referred to as a broken line). The apices of this broken line are the points $A_1 \ldots, A_n$. (In what follows we shall understand by "points of the broken line" only the points of its apices.) We shall be interested in local homogeneous rules of movement of apices that will cause all apices to lie on a straight line, the distances between apices A_i and A_{i+1} $(i = 1, \ldots n - 1)$ being equal to a present value d; in addition, point A_i lies between points A_{i-1} and A_{i+1}. In this state a broken line will be called stationary. The term *local homogeneous rule* means that the movement of each apex depends only on a small number of neighbors (localness) and further that this dependence is the same for all internal apices (homogeneity). We assume time to be discrete, that is, the movements of apices of the broken line take place stepwise. It would also be possible to examine the analogous problem in continuous time. Such a problem would be described by a system of differential equations, but evidently this method would not lead to significant differences in principle.

We shall assume that the motility rules for the broken line have the following properties:

1. Translocation of apices with respect to their neighbors depends only on the positions of apices relative to their neighbors and does not depend on their position in the plane.

2. The broken line does not move when and only when it is in the stationary state.

Now we shall examine the simplest case in which the translocation of an apex of the broken line depends only on itself and its two nearest neighbors to the left and to the right. The translocation of a terminal apex is defined by its position and that of its neighbor. Thus the rules of translocation of the broken line are defined by three vectorial functions, f, ϕ_1, and ϕ_2:

$$A_i(t + 1) = f[A_{i-1}(t), A_i(t), A_{i+1}(t)] \qquad (i = 2, 3, \ldots, n - 1),$$

$$A_1(t + 1) = \phi_1[A_1(t), A_2(t)], \qquad A_n(t + 1) = \phi_2[A_n(t), A_{n-1}(t)].$$

We shall assume that $\phi_1 = \phi_2 = \phi$. Because the rules are subject to conditions 1 and 2 functions f and ϕ cannot be arbitrary. Condition 1 leads to the following relations:

$$f(r_{-1} + \rho, r_0 + \rho, r_1 + \rho) = f(r_{-1}, r_0, r_1) + \rho \tag{1}$$

$$f(Ar_{-1}, Ar_0, Ar_1) = Af(r_{-1}, r_0, r_1), \tag{2}$$

$$\phi(r_0 + \rho, r_1 + \rho) = \phi(r_0, r_1) + \rho, \qquad \phi(Ar_0, Ar_1) = A\phi(r_0, r_1),$$

where r_{-1}, r_0, r_1, and ρ are any vectors of the plane and A is an arbitrary rotation. From equation (2) we obtain $\phi(r_0, r_1) = r_0$ when and only when the distance between the points r_0 and r_1 is equal to d: $|r_0 - r_1| = d$. From the same condition it follows that if points r_{-1}, r_0, and r_1 are located on a single line and point r_0 is located between points r_{-1} and r_1 at a distance d from each of them, then $f(r_{-1}, r_0, r_1) = r_0$. Conversely, if $f(r_{-1}, r_0, r_1) = r_0$, then points r_{-1}, r_0, and r_1 are located on the same line and point r_0 is located between points r_{-1} and r_1 at the same distance (not necessarily equal to d) from each of them. Thus, if

$$r_0 = \tfrac{1}{2}(r_{-1} + r_1), \quad |r_{-1} - r_0| = |r_1 - r_0| = d, \quad \text{then} \quad f(r_{-1}, r_0, r_1) = r_0,$$

$$\tag{3}$$

and if

$$f(r_{-1}, r_0, r_i) = r_0, \quad \text{then} \quad r_0 = \tfrac{1}{2}(r_{-1} + r_1).$$

The following rules for the translocation of the broken line were used for modeling on the computer. For internal points the point $A_i(t + 1)(1 < t < n)$ is defined as the weighted mean of points $A_i(t)$, A_i', A_i'': $A_i(t + 1) = \sigma_1 A_i' + \sigma_2 A_i'' + (1 - \sigma_1 - \sigma_2)A_i(t)$, where $A_i' = \tfrac{1}{2}[A_{i-1}(t) + A_{i+1}(t)]$ is the mean of points $A_{i-1}(t)$ and $A_{i+1}(t)$, and A_i'' is any point whose distance to points $A_{i-1}(t)$ and $A_{i+1}(t)$ is d and whose position is on the same side of line $A_{i-1}(t)$ $A_{i+1}(t)$ as point $A_i(t)$, if such a point exists [this will occur when the distance between points $A_{i-1}(t)$ and $A_{i+1}(t)$ is not greater than $2d$], or if such a point does not exist, $A_i'' = A_i'$. For terminal points $A_1(t)$ and $A_n(t)$, the rule is that point $A_2(t + 1)$ is the weighted mean of points $A_1(t)$ and A_1':

$$A_1(t + 1) = \gamma A_1' + (1 - \gamma)A_1(t),$$

where A_1' is the point lying on the line $A_1(t)A_2(t)$ at a distance d from the point $A_2(t)$ and on the same side of point $A_2(t)$ as the point $A_1(t)$. The rule is the same for point $A_n(t)$, but instead of points $A_1(t)$, $A_1(t + 1)$, $A_2(t)$, and A_1', the points $A_n(t)$, $A_n(t + 1)$, $A_{n-1}(t)$, and A_n' are to be taken. It is easy to calculate that for this rule, when $|r_{-1} - r_1| \leqslant 2d$,

$$f(r_{-1}, r_0, r_1) = \sigma_1 \frac{r_{-1} + r_i}{2} + \sigma_2 \psi(r_{-1}, r_0, r_1) + (1 - \sigma_1 - \sigma_2)r_0, \quad (4)$$

where $\psi(r_{-1}, r_0, r_1)$ denotes the point in the plane from which the distance to points r_{-1} and r_1 is equal to d and that lies on the same side of the line passing through r_{-1} and r_1, as does point r_0. If $|r_{-1} - r_1| \leqslant 2d$, then $\psi(r_{-1}, r_0, r_1) = (r_{-1} + r_2)/2$;

$$\phi(r_0, r_i) = \gamma\left(\frac{r_0 - r_1}{|r_0 - r_1|}d + r_1\right) + (1 - \gamma)r_0. \quad (5)$$

Clearly, it is not permissible to set $\sigma_1 = 0$, because then any position of the broken line in which the distance between neighboring points is equal to d will be immobile, and no approach to coincidence with the stationary state will occur. From the same considerations it is not possible to set $\gamma = 0$.

For the case $\sigma_2 = 0$, the function f has the form

$$f(r_{-1}, r_0, r_1) = \sigma\frac{r_{-1} + r_1}{2} + (1 - \sigma)r_0, \quad (6)$$

where instead of σ_1 we write σ. In this case f is a linear function. It is easy to see that if points r_{-1}, r_0, and r_1 lie almost on a line and the distance between them is close to d, then to an accuracy of the first order in the above terms the function will have, instead of its form in equation (4), the form it has in equation (6), where $\sigma = \sigma_1 + \sigma_2$.

The problem would be greatly simplified if, for a stationary state, we did not require that the distances between points be a preset value d but that they simply be equal. For such a problem there exist simple and easily studied linear rules of translocation. When there is a preset distance between successive apices, linear rules, as is easy to see, do not exist. Nevertheless, this problem is of interest both because it represents a biologically more natural situation and because it models difficulties that arise in analogous problems with a more complex stationary state.

We shall now show that natural assumptions lead to function f of equation (6). First we shall find what kind of linear functions can exist, that is, functions f of the form

$$f(r_{-1}, r_0, r_1) = B_{-1}r_{-1} + B_0r_0 + B_1r_1 + b,$$

where B_{-1}, B_0, and B_1 are certain matrices of the second order and b is a certain vector in the plane. From equation (1) we find that $B_{-1} + B_0 + B_1 = E$, and from equation (2) that $b = 0$ and that each of the matrices B_{-1}, B_0, and B_1 has the form $B_j = \alpha_j E + \beta_j I$ ($j = -1, 0, 1$), where

$$I \begin{pmatrix} 0 & -1 \\ 1 & 0 \end{pmatrix}.$$

From equation (3) we find

$$\alpha_1 = \alpha_{-1} = \tfrac{1}{2}(1 - \alpha_0), \qquad \beta_1 = \beta_{-1} = -\tfrac{1}{2}\beta_0.$$

Thus we have found all linear functions of f.

If one makes the still quite natural assumption that function f is invariant not only with respect to rotation but also with respect to arbitrary reflection, then $\beta_{-1} = \beta_0 = \beta_1 = 0$, and assuming $\alpha_0 = 1 - \sigma$ we arrive at function f in equation (6). The requirement of invariance of rules with respect to reflection is related to the fact that the problem of line straightening is confined to a plane. For analogous problems in three dimensions any linear rule leads to function f in equation (6).

We have modeled on a computer the translocation process of the broken line described by functions f and ϕ in equations (4) and (5). The "spiral," "hook," "harmonica," and "corner" were taken as initial configurations (figs. 3–8 and tables 1–5). It was found that even for such initial configurations as the spiral, which differ so much from the stationary state, movement toward the stationary state occurs. Nevertheless, it would be incorrect to affirm that any initial configurations move toward the stationary state.

For example, let us examine a broken line with three apices A_1, A_2, and A_3, where the apices A_1 and A_3 coincide. It is clear that in the course of movement the points A_1 and A_3 will always remain in coincidence. It is possible to show that if the distance between the points A_1 and A_2 is equal to $[\gamma/(\sigma + \gamma)]d$, the translocation of the broken line will take place as a parallel translation. It is also possible to show that for certain initial configurations of the broken line close to those mentioned above, there will likewise be no movement toward the stationary state.* These examples of initial configurations are, of course, "pathological." It is very probable that for all nonpathological initial states, there is movement toward the stationary state. The term *pathological*, it would appear, refers to a configuration of the broken line all of whose apices lie on a single line and at the same time not all of whose internal apices occur within a cut passing through the terminal apices. An example of a pathological configuration would be the above-mentioned broken line with three apices. It should be understood that we do not affirm that there is no movement toward the stationary state starting with any pathological configuration. Our hypothesis is that there will be movement toward the stationary state starting from nonpathological initial states.

As a measure of the deviation from the stationary state of the broken line with apices $r_1 \ldots r_n$, the following quantities were used:[†]

$$h = \max_{1 < i < n} \left| r_i - \frac{r_{i-1} + r_{i+1}}{2} \right| \quad \text{and} \quad \delta = \max_{1 < i < n} |d - |r_{i+1} - r_i||,$$

where $|\rho|$ is the length of the vector ρ.

In all cases examined by modeling, the quantities h and δ, starting with some t_0, monotonically decreased; this process began with configurations still quite far from the stationary state. In the tables a note is made of time t_0. As might have been expected, it is small for simple geometrical configurations such as the hook and large for complex ones such as the spiral.

From the results of modeling we concluded that in the beginning (rough straightening), the quantities h and δ decrease more rapidly than at the end. It is also of interest that the increase of time due to the increase in the number of apices for rough straightening is proportionately less.

Rather unexpected is the strong dependence of the rate of convergence toward the stationary state on the value of σ_2. Convergence is significantly slower for $\sigma_2 > 0$ than for $\sigma_2 = 0$. This apparently indicates that it is more efficient to first align on a line and then move to the correct distances than it is to do this in reverse order. When $\sigma_2 > 0$, each cell from the very start is guided in its movements by "knowledge" of the distance d. In the given

---

* One has in mind here the case where the points A_1, A_2, and A_3 lie on a line, points A_1 and A_3 are near each other, and the distance between points A_1 and A_2 is close to $[\gamma/(\sigma + \gamma)]d$.

† Another measure of deviation that can be used is H, the maximum distance of apices from a line connecting the ends of the broken line. It is possible to show that

$$H < \frac{d + \delta}{d - \delta} h \frac{n^2}{4};$$

therefore if $h \to 0$ when $t \to \infty$, then also $H \to 0$.

case it appears that this knowledge only slows down the process of line straightening and consequently is superfluous. Compared with the rate of convergence toward the stationary state, the dependence of the position of the limit line on the parameters σ_1, σ_2, and γ is weaker.

It is evident that if the initial configuration of the broken line has a center or axis of symmetry, the final position has it also. Thus if the initial state of the broken line has a center of symmetry, the limit line passes through it; if the initial state has an axis of symmetry, the limit line is perpendicular to it.

In the general case the position of the limit line is difficult to predict. We note that when the initial broken line does not have a center of symmetry, the limit line may not even intersect the convex enveloping curve passing through all initial positions of the apices (cf. the spiral and corner).

Now we shall discuss the results for each type of initial configuration separately.

Hook (fig. 3, table 1). In the limit state, point r_1 rose slightly, whereas point r_n scarcely moved at all. The position of point r_1 depends basically

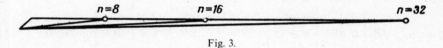

Fig. 3.

Table 1.

№	n	σ_1	σ_2	γ	t_0	$t=10$		$t=40$		$t=80$	
						h	δ	h	δ	h	δ
1	8	$\frac{1}{2}$	0	$\frac{1}{2}$	10	0,0406	0,0167	0,0072	0,0062	0,0009	0,0009
2	16	$\frac{1}{2}$	0	$\frac{1}{2}$	10	0,0406	0,0167	0,0103	0,0085	0,0052	0,0052
3	32	$\frac{1}{2}$	0	$\frac{1}{2}$	10	0,0406	0,0167	0,0103	0,0085	0,0052	0,0052
4	8	$\frac{1}{4}$	$\frac{1}{4}$	$\frac{1}{2}$	10	0,0754	0,0063	0,0532	0,0034	0,0447	0,0025
5	16	$\frac{1}{4}$	$\frac{1}{4}$	$\frac{1}{2}$	80	0,0754	0,0063	0,0536	0,0035	0,0472	0,0029
6	32	$\frac{1}{4}$	$\frac{1}{4}$	$\frac{1}{2}$	170	0,0754	0,0063	0,0536	0,0035	0,0472	0,0029

continued

Table 1 cont.

$t = 160$		$t = 320$		$t = 480$		$A_1(T)$		$A_n(T)$		T
h	δ	h	δ	h	δ	x_1	y_1	x_n	y'_n	
						—0,9	—0,86	6,038	0,003	120
0,0021	0,0025					—0,9	—0,86	4,024	0	240
0,0026	0,0031	0,0013	0,0018	0,0008	0,0012	—0,9	—0,86	30,007	0	480
						—0,88	—0,86	6,021	0	120
0,0403	0,0021	0,0300	0,0012	0,0222	0,0007	—0,9	—0,87	14,021	0	480
0,0403	0,0021	0,0316	0,0013	0,027	0,0010	—0,9	—0,87	30,004	0	720

on the length of segment r_1r_2 and on angle $\angle r_1r_2r_3$ and is almost independent of the number of apices n. As might have been expected, the rate of line straightening is also weakly dependent on the value of n.

Centrosymmetrical harmonica (figs. 4 and 5, table 2). In this case the limit line passes through the center of symmetry. If the number of apices n is large, the line joining the ends of the harmonica tends toward the axis x.

Harmonica with an axis of symmetry (fig. 6, table 3). In this case the limit position of the broken line must be perpendicular to the axis of symmetry. It is of interest that if $\sigma = \gamma \leqslant \frac{1}{2}$, the limit position is always below the line r_1r_n; for $\sigma = \gamma = 1$ an oscillation occurs, causing the limiting configuration to be somewhat above line r_1r_n (fig. 6).

Corner (fig. 7, table 4). In this case, as in the preceding one, the initial configuration has an axis of symmetry, and therefore the limiting configura-

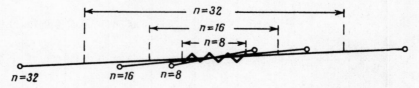

Fig. 4.

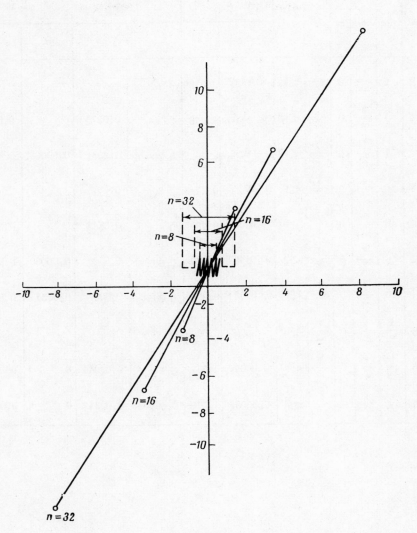

Fig. 5.

Table 2.

№	n	$tg\,\varphi$	σ_1	σ_2	γ	t_0	$t=40$		$t=240$		$t=480$	
							h	δ	h	δ	h	δ
1	8	1	$\frac{1}{2}$	0	$\frac{1}{2}$	<40	0,00005	0,0339				
2	16	1	$\frac{1}{2}$	0	$\frac{1}{2}$	<40	0,0055	0,246	0	0,0278	0	0,0020
3	32	1	$\frac{1}{2}$	0	$\frac{1}{2}$	<40	0,0066	0,2927	0,00076	0,2034	0,000073	0,1101
4	8	10	$\frac{1}{2}$	0	$\frac{1}{2}$	<40	0,008	0,1716	0	0		
5	16	10	$\frac{1}{2}$	0	$\frac{1}{2}$	<40	0,0100	0,782	0	0,0885	0	0,0063
6	32	10	$\frac{1}{2}$	0	$\frac{1}{2}$	<40	0,0145	0,900	0,0025	0,652	0,00030	0,354
7	64	10	$\frac{1}{2}$	0	$\frac{1}{2}$	<40	0,0145	0,900	0,0054	0,898	0,0034	0,851
8	8	$\frac{1}{10}$	$\frac{1}{2}$	0	$\frac{1}{2}$	<40	0	0,0009	0	0		
9	16	$\frac{1}{10}$	$\frac{1}{2}$	0	$\frac{1}{2}$	<40	0,00050	0,0041	0	0,00046	0	0000033
10	32	$\frac{1}{10}$	$\frac{1}{2}$	0	$\frac{1}{2}$	<40	0,00062	0,0049	0,00006	0,0034	0	0,0018

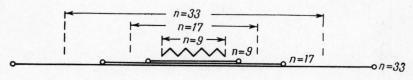

Fig. 6.

Table 2 cont.

$t = 960$		$t = 1920$		$t = 2880$		$A_1(T)$		$A_n(T)$		T
h	δ	h	δ	h	δ	x_1	y_1	x_n	y_n	
						—3,44	—0,597	3,44	0,597	120
						—7,457	—0,703	+7,457	0,703	480
0	0,0321	0	0,00272			—15,451	—0,820	+15,451	0,820	1920
						—1,271	—3,261	+1,271	3,261	960
0	0,00003					—3,40	—6,68	3,40	6,68	960
0	0,1031	0	0,0087	0	0,00074	—8,23	—13,12	8,23	13,12	2880
0,0012	0,659	0,00016	0,364	0,00002	0,2007	—18,14	—24,24	18,14	24,24	4800
						—3,5	—0,050	3,5	0,050	240
0	0					—7,50	—0,050	7,50	0,050	960
0	0,00054					—15,49	—0,050	15,49	0,050	960

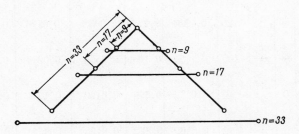

Fig. 7.

Table 3.

№	n	tgφ	σ_1	σ_2	γ	$t_{\bullet}$	$t=40$		$t=120$		$t=240$	
							h	δ	h	δ	h	δ
1	9	1	$\frac{1}{2}$	0	$\frac{1}{2}$	<10	0,0143	0,095	0,0007	0,0045		
2	17	1	$\frac{1}{2}$	0	$\frac{1}{2}$	<40	0,0079	0,257	0,0055	0,124	0,00020	0,0396
3	33	1	$\frac{1}{2}$	0	$\frac{1}{2}$	<40	0,0066	0,292	0,0025	0,272	0,0023	0,211
4	17		1	0	1	40	0,423	0,108	0,089	0,0357	0,0086	0,0038
5	17		$\frac{1}{4}$	$\frac{1}{4}$	$\frac{1}{2}$	40	0,350	0,196	0,285	0,122	0,193	0,055

Table 4.

№	n	d_0	σ_1	σ_2	γ	$t_{\bullet}$	$t=40$		$t=240$		$t=480$	
							h	δ	h	δ	h	δ
1	9	1	$\frac{1}{2}$	0	$\frac{1}{2}$	<40	0,0525	0,1911	0,00003	0,0001	0	0
2	17	1	$\frac{1}{2}$	0	$\frac{1}{2}$	<20	0,0636	0,2896	0,0142	0,1070	0,0016	0,0124
3	33	1	$\frac{1}{2}$	0	$\frac{1}{2}$	80	0,0636	0,290	0,0257	0,2875	0,0168	0,2373
4	17	1	1	0	1	20	0,0864	0,2792	0,0032	0,0124	0,00003	0,00012
5	17	1	$\frac{1}{4}$	$\frac{1}{4}$	$\frac{1}{2}$	100	0,2852	0,1082	0,2534	0,0911	0,2142	0,0649
6	17	$\frac{1}{2}$	$\frac{1}{2}$	0	$\frac{1}{2}$	<40	0,0142	0,593	0,0193	0,177	0,0025	0,021
7	33	$\frac{1}{4}$	$\frac{1}{2}$	0	$\frac{1}{2}$	440	0,0160	0,822	0,0195	0,672	0,0195	0,480

Table 3 cont.

$t = 480$		$t = 960$		$t = 1920$		$A_1(T)$		$A_n(T)$		T
h	δ	h	δ	h	δ	x_1	y_1	x_n	v_n	
						—4	—0,23	4	—0,23	160
0,00021	0,0039	0	0,00008			—8	—0,32	8	—0,32	960
0,0016	0,119	0,0006	0,0378	0,00006	0,0037	—16	—0,42	16	—0,42	2880
0,00008	0,00004	0	0			—8	+0,2	8	+0,2	960
0,084	0,0102	0,015	0,0003	0,0026	0,00001	—8	—0,72	8	—0,72	1920

Table 4 cont.

$t = 960$		$t = 1920$		$t = 2880$		$A_1(T)$		$A_n(T)$		T
h	δ	h	δ	h	δ	$x_1(T)$	$y_1(T)$	$x_n(T)$	$v_n(T)$	
0	0					—4	—0,425	4	—0,425	960
0,00002	0,00012					—8	—0,852	8	—0,852	960
0,0070	0,1068	0,0008	0,0124	0,00008	0,00125	—15,99	—1,708	15,99	—1,708	2880
0	0					—8	—0,844	8	—0,844	960
0,0943	0,0122	0,0597	0,0048	0,0233	0,0007	—7,98	—1,45	7,98	—1,45	2880
0,00002	0,0002					—8	—3,017	8	—3,017	960
0,0114	0,224	0,0016	0,028			—15,72	—8,55	1572	—8,55	1920

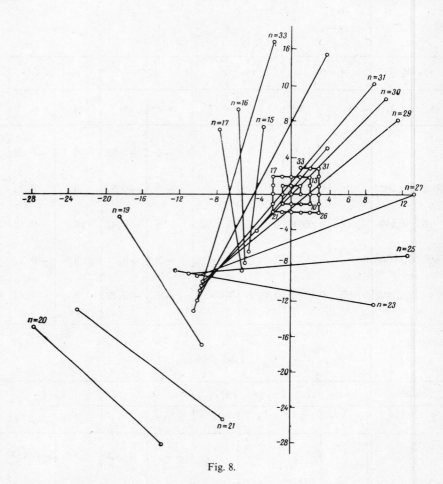

Fig. 8.

tion must be perpendicular to the axis of symmetry. The computer experiments suggest that the limit line always descends below line $r_1 r_n$ (at least this is so if the distances between the apices in the initial configuration are not greater than d) and that the value of this translocation c is proportional to n for a given angle and distances between points. Evidently, the coefficient of proportionality is greater the smaller the angle and the less the distances between successive apices. When the angle is $\pi/2$ and $d = 1$, the empirically derived formula has the form $c = 0.053(n - 1)$.

Spiral (fig. 8, table 5). We modeled the geometrical linearization of spirals with various numbers of segments. The results obtained suggest that the position angles of the limit configurations of the broken line, given spirals of varying length, are distributed approximately proportionately to these lengths.

Table 5.

Nº	n	d_e	σ_1	σ_2	γ	t_0	$t = 40$		$t = 240$		$t = 480$	
							h	δ	h	δ	h	δ
1	33	1	$\frac{1}{2}$	0	$\frac{1}{2}$	480	0,0634	0,8761	0,0113	0,762	0,0136	0,476
2	33	1	1	0	1	40	0,1019	0,9943	0,0481	0,5008	0,0201	0,2109
3	33	1	$\frac{1}{4}$	$\frac{1}{4}$	$\frac{1}{2}$	1240	0,4836	0,3229	0,4660	0,3588	0,4118	0,2654
4	33	3	$\frac{1}{2}$	0	$\frac{1}{2}$	520	0,1442	0,8343	0,0196	0,7801	0,0187	0,5210

Table 5 cont.

$t = 960$		$t = 1920$		$t = 2880$		$A_n(T)$		$A_1(T)$		T
h	δ	h	δ	h	δ	x_n	y_n	x_1	y_1	
0,0079	0,184	0,0010	0,0211			—1,96	17,1	—10,5	—13,2	1920
0,0024	0,0585	0	0,0397			—1,75	16,9	—9,19	—14,1	1920
0,5336	0,4664	0,2987	0,1373	0,1933	0,0558	12,15	2,45	—2,85	2,91	2880
0,0114	0,2349	0,0017	0,0299	0,0002	0,0030	10,67	17,17	11,14	—14,76	2880

Thus the computer experiments show that approach to the stationary state takes place for certain values of the parameters σ and γ.

In concluding the authors would like to note the very valuable discussions of various problems in embryology with A. A. Neifakh and L. V. Belousov. Without these discussions the present work would not have been possible.

REFERENCES

1. Von Neumann, J. (1966) *Theory of self-reproducing automata*. Edited and completed by Arthur W. Burks. Urbana: University of Illinois Press.
2. Thom, R. (1969) Stabilité structurelle et morphogenése. Menlo Park, Calif.: W. A. Benjamin.
3. Levenshtein, V. I. (1965) On one method of solving problems of synchronizing chains of automata in minimal time. *Probl. Peredachi Inform.* (*Probl. Inform. Transfer*) 1, no. 4.
4. Levenshtein, V. I. (1968) On the synchronization of two-directional nets of automata. *Probl. Peredachi Inform.* (*Probl. Inform. Transfer*) 4, no. 4.
5. Gurvich, A. G. (1944) The theory of the biological field. *Sov. Nauk.* (*Sov. Sci.*).

6. Waddington, C. H. (1940) *Organisers and Genes.* Cambridge, Eng.: Cambridge University Press, p. 109.
7. Saxen, L. and Toivonen, S. (1962) *Primary Embryonic Induction.* London: Logos Press; New York: Academic Press.
8. Spemann, H. and Mangold, H. (1924) Über Induction von Embryoanlagen durch Implantation Artfremde Organizatoren. *Wilhelm Rous's Archiv Microscop. Anat. Entwicklungsmech.* 100: 599–638.
9. Bonner, J. F. (1965) *The Molecular Biology of Development.* Oxford University Press: New York.
10. Moscona, M. H. and Moscona, A. A. (1963) Inhibition of adhesiveness and aggregation of dissociated cells by inhibition of protein and RNA synthesis. *Science* 142: 1070–71.
11. Tsetlin, M. L. (1963) Finite automata and the modeling of the simplest forms of behavior. *Results Math. Sci.* 18, no. 4.
12. Ebert, J. D. (1965) *Interacting Systems in Development.* Holt, Rinehart and Winston: New York.

2. The Problem of Circularization in Mathematical Modeling of Morphogenesis

A. M. Leontovich, I. I. Pyatetskii-Shapiro,
and O. N. Stavskaya*

Local homogeneous motility rules of cells that lead to circularization are examined, and results of computer studies are presented. It is shown that circularization takes place for a broad class of initial configurations. For the linear rule, convergence to the circle is proved.

Introduction

We have shown that the data of experimental embryology suggest that processes of morphogenesis occur because each cell has the ability to move in a direction that depends on a small number of its neighbors.[1] A detailed consideration of the hypothesis in reference 1 has made it possible to characterize a class of mathematical problems concerned with modeling such processes. Such problems reduce to the investigation of local homogeneous rules of movement of separate elements—the cells—which lead to the appearance of a tissue of a given geometrical form. The simplest mathematical problem based on this hypothesis, that of straightening a line, was examined in reference 1. The present paper considers the problem of circularization.

We examine here local homogeneous rules of movement for the apices of a polygon that cause the polygon to tend to become regular. Section 1 describes the results of computer modeling achieved by using certain motility rules. It is found that for suitable values of parameters and with not excessively bad initial configurations of the apices of the polygon, there is convergence toward a regular polygon (although for some initial configurations there is no convergence). An instructive example of motility rules that seem entirely natural and yet do not produce convergence from almost any initial configuration is given at the end of Section 1. In Section 2 one of the simpler motility rules, a linear one, is studied in more detail.

1. Statement of the Problem: results of Computer Modeling

We examine a closed broken line lying in a plane and having n apices $A_1, \ldots, A_n$. We shall be interested in the local homogeneous rules of motion

* First published in *Avtomatika i Telemekhanika* (1971) 2: 100–10.

of the apices that will cause the broken line to converge towards the configuration of a regular polygon. We shall call this problem circularization. The term *local homogeneous rules* means that the movement of each apex depends only on a small number of neighbors (localness)—that is, on itself and on k neighbors to the right and k neighbors to the left—and that this dependence is the same for all apices (homogeneity). Time is taken to be discrete, that is, movement takes place stepwise. The rules of motion are therefore expressed by a vectorial function f:*

$$A_j(t + 1) = f(A_{j-k}(t), \ldots, A_{j-1}(t), A_j(t), A_{j+1}(t), \ldots, A_{j+k}(t)).$$

We shall assume that translocation of an apex relative to its neighbors depends only on the mutual distribution of apices and on its k neighbors to the right and k neighbors to the left, and does not depend on its position on the plane. This leads to a constraint being placed on function f: no matter what the vectors $r_{-k}, \ldots, r_{-1}, r_0, \ldots, r_k$, ρ and the operator of rotation A are, the following relations hold:

$$f(r_{-k} + \rho, \ldots, r_0 + \rho, \ldots, r_k + \rho) = f(r_{-k}, \ldots, r_0, \ldots, r_k) + \rho \qquad (1)$$

$$f(Ar_{-k}, \ldots, Ar_0, \ldots, Ar_k) = Af(r_{-k}, \ldots, r_0, \ldots, r_k). \qquad (2)$$

It is clear that given this definition of the motility rule of apices, a closed broken line all of whose segments are equal and all of whose angles are the same (such a broken line will be called regular) transforms into a line similar to itself. It is easy to describe such a broken line with n apices, which lie on a circle circumscribing a regular polygon (we shall assume that the numeration of the apices of this polygon proceeds clockwise). However, the numeration differs from the usual case. Between neighboring apices of the broken line the regular polygon has $s - 1$ apices, where s may be equal to $1, 2, \ldots, n - 1, n$. We shall call such broken lines regular and of order s. It is clear that all regular broken lines of a given order s are similar to one another. When $s = n$, all the apices of the broken line merge into one; when $s = 1$ and $s = n - 1$, the broken line is a regular polygon. In addition, when $s = 1$ the numeration of apices is clockwise; when $s = n - 1$ it is counterclockwise. When $1 < s < n - 1$, the broken line is self-intersecting.

In the discussion below, the function f will be assumed to be such that the only stationary polygons (that is, all of those whose apices do not move) are regular broken lines. Here one may discern two cases. For a problem of the first type the stationary polygons are all possible regular broken lines of order 1, whereas for a problem of the second type they are only those with a fixed length of segment d. Function f can be such that for all values of n, a regular n-segment broken line of order 1 will be stationary (also for, a

---

* Function f may be undefined only in those cases where one or more of the points coincide with others.

problem of the second type, d may depend on n); furthermore, it may be such that this occurs for only one value of n.

The following functions f were examined. We note that in the rules considered here, movement must be such that the angle between the vectors $r_1 - r_{-1}$ and $f - \frac{1}{2}(r_1 r_{-1})$ is equal to $\pi/2$ and not to $3\pi/2$ (here and in subsequent cases all angles are measured counterclockwise). Taking into account this remark, the functions f that will be presented below are uniquely defined as follows:

1. $f = f_1(r_{-1}, r_0, r_1)$ is defined by the condition

$$|f_1 - r_1| = |f_1 - r_{-1}| = d.$$

2. $f = f_{2,n}(r_{-1}, r_0, r_1)$ is defined by the condition

$$\angle r_{-1} f_{2,n} r_1 = r_1 = \pi - \frac{2\pi}{n} \quad \text{and} \quad |f_{2,n} - r_1| = |f_{2,n} - r_{-1}|$$

(thus $f_{2,n}$ depends on n).

3. $f = f_3(r_{-2}, r_{-1}, r_0, r_1, r_2)$ is defined by the condition

$$\cos \angle r_{-1} f_3 r_1 = \frac{1 - \gamma}{2} (\cos \angle r_{-2} r_{-1} r_0 + \cos \angle r_0 r_1 r_2)$$

$$+ \gamma(\cos \angle r_0 r_1 r_2)$$

and

$$|f_3 - r_1| = |f_3 - r_{-1}|.$$

4. $f = f_4(r_{-2}, r_{-1}, r_0, r_1, r_2) = \frac{1}{2}f_4^+(r_{-2}, r_{-1}, r_0, r_1, r_2) + \frac{1}{2}f_4^-(r_{-2}, r_{-1}, r_0, r_1, r_2)$, where the point $f_4^\pm$ is the intersection of a line perpendicular to the vector $r_1 - r_{-1}$ and passing through the point $\frac{1}{2}(r_1 + r_{-1})$ (therefore $|f_4^\pm - r_1| = |f_4^\pm - r_{-1}|$) and of the circle passing through the point $r_1, r_{-1}, r_{\pm 2}$.

5. $f = f_5(r_{-1}, r_0, r_1)$ is defined by the condition that the points $r_{-1}, f_5,$ and r_1 must lie on a circle of a given radius R if $R \geqslant \frac{1}{2}|r_{-1} - r_1|$, and on a circle of radius $\frac{1}{2}|r_{-1} - r_1|$ if $R \leqslant \frac{1}{2}|r_{-1} r_1|$.

Taking the weighted mean of the functions listed above and also of the function $f = r_0$, we obtain problems of circularization with different rules of motility for the apices. On the computer we modeled problems with the following functions f:

$$f = (1 - \sigma)r_0 + \sigma f_4 \tag{3}$$

$$f = (1 - \sigma_1 - \sigma_2)r_0 + \sigma_1 f_1 + \sigma_2 f_3, \qquad \sigma_2 \neq 0, \tag{4}$$

$$f = (1 - \sigma_1 - \sigma_2)r_0 + \sigma_1 f_1 + \sigma_2 f_{2,n}, \qquad \sigma \neq 0. \tag{5}$$

Function f from equation (3), function f from (4) when $\sigma_1 = 0$, and function f from (5) when $\sigma_1 = 1$ are examples of problems of circularization of the second type. Furthermore, in equations (3) and (4) the number of apices of the polygon can be arbitrary, whereas in equation (5) it is equal to the given number n.

Let us note that if in equation (5) $\sigma_1 = 0$, f has the form

$$f = (1 - \sigma)r_0 + \sigma f_{2,n}. \tag{6}$$

This is a linear function. We shall show that natural assumptions lead to a function of the form that occurs in equation (6). In other words, we shall assume that $k = 1$; that function f is linear:

$$f(r_{-1}, r_0, r_1) = B_{-1}r_{-1} + B_0 r_0 + B_1 r_1 + b;$$

and that the dependence on the right neighbor is the same as the dependence on the left (that is, the discussion concerns the symmetrical case). Let the number of apices of the polygon be n. From equation (1) we obtain

$$B_{-1} + B_0 + B_1 = E, \tag{7}$$

and from equation (2) it follows that $b = 0$ and B_{-1}, B_0, and B_1 are transposable by any rotation. From this it follows that each of the matrices B_{-1}, B_0, and B_1 can be put in the form

$$B_j = \alpha_j E + \beta_j I, \qquad j = -1, 0, 1,$$

where $I = \begin{pmatrix} 0 & -1 \\ 1 & 0 \end{pmatrix}$, and on the basis of equation (7)

$$\alpha_{-1} + \alpha_0 + \alpha_1 = 1, \qquad \beta_{-1} + \beta_0 + \beta_1 = 0.$$

From the fact that stationary polygons are regular (their sides and angles are equal), it is not difficult to obtain

$$\alpha_{\pm 1} = \tfrac{1}{2}\left(1 - \alpha_0 \pm \beta_0 \operatorname{tg}\frac{\pi}{n}\right), \qquad \beta_{\pm 1} = \tfrac{1}{2}\left[\pm(1 - \alpha_0) \operatorname{tg}\frac{\pi}{n} - \beta_0\right],$$

where α_0 and β_0 can be arbitrary. From this we obtain

$$f(r_{-1}, r_0, r_1 = r_0 - (1 - \alpha_0)[r_0 - \tfrac{1}{2}(r_1 + r_{-1})]$$

$$+ (1 - \alpha_0)\tfrac{1}{2}\operatorname{tg}\frac{\pi}{n}I(r_1 - r_{-1}) + \beta_0 I[r_0 - \tfrac{1}{2}(r_1 + r_{-1})]$$

$$+ \beta_0 \tfrac{1}{2}\operatorname{tg}\frac{\pi}{n}(r_1 - r_{-1}). \tag{8}$$

Thus we have found the general form of function f for the linear case. Because the discussion concerns the symmetrical case, it is easy to see that $\alpha_{-1} = \alpha_1, \beta_{-1} = -\beta_1, \beta_0 = 0$. Setting $\alpha_0 = 1 - \sigma$, we obtain

$$f(r_{-1}, r_0, r_1) = (1 - \sigma)r_0 + \sigma\left[\tfrac{1}{2}(r_1 + r_{-1}) + \tfrac{1}{2}\operatorname{tg}\frac{\pi}{n}I(r_1 - r_{-1})\right]. \tag{9}$$

When $\sigma = 1$, the function obtained coincides with $f_{2,n}$, and for an arbitrary σ, the function f coincides with function f from equation (6).

It is apparent that all our proposed rules are not invariant with respect to reflection. In addition, it is evident from equation (8) that there are no linear rules invariant with respect to reflection. It is possible to show that for the analogous problem of circularization in three dimensions there are no invariant linear rules even with respect to rotation.*

The results of modeling are shown in tables 1–5. As a measure of the deviation of the closed broken line from a regular polygon, the quantity $\Delta_1 = (R - R_{min})/R$ is taken, where $R_{min} = \min_{1 \leqslant i \leqslant n} R_i$, $R_i = R_i(t)$ is the radius of a circle passing through points A_{i-1}, A_i, and A_{i+1} (thus if $R = $ constant, R_i is the radius of a circle circumscribing the broken line) and R is the radius of a circle circumscribing the limiting state of the polygon [that is, $R - \lim_{t \to \infty} R_i(t)$ and the value of R is found from the results of modeling]. When the stationary broken line is a regular polygon with sides of a given length (i.e., in problems of the second and fourth type), the quantity $\Delta_2 = \max_{1 \leqslant i \leqslant n} |d - |A_i - A_{i+1}||$. As initial conditions for modeling functions f in equations (3) and (5), an equilateral triangle was used, on each of whose sides, at a distance of 1 from each other, an equal number of points were marked off. For function f in equation (4), a "flower" (see fig. 1) and a rectangle were also taken.

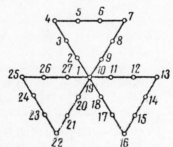

Fig. 1. Flower.

Table 1 shows the results of the modeling of the function in equation (5). As can be seen, in the beginning circularization proceeds faster (with $\sigma_1 \neq 0$). We note that this conclusion also holds for other functions f. In this respect it is useful to compare the results with $\sigma_1 \neq 0$ and $\sigma_1 = 0$ when function f is linear and circularization proceeds at a constant rate. From the data shown in table 1 it is seen that when $\sigma_1 \neq 0$, the initial circularization proceeds more slowly than when $\sigma_1 = 0$ but later significantly faster. We see that for

* This is a reasonable supposition if any regular polygon must be a stationary broken line. If it be required only that regular polygons lying in planes parallel to a given plane be stationary, linear rules exist.

Table 1. Table of values $\Delta_1(t)$, $\Delta_2(t)$, and $\rho_{max} = \max_i |A_i - A_{i+1}|$, $\rho_{min} = \min_i |A_i - A_{i+1}|$ for f in equation (5) when $d = \frac{4}{3}$ and initial configuration on an equilateral triangle.

n	σ_1	σ_2	t	15	30	75	150	300	600	1200
12 × 3	0	0,5	Δ_1	—	—	0,0150	0,0005	0,0000	—	—
			ρ_{max}	—	—	0,8404	0,8284	0,8280	—	—
			ρ_{min}	—	—	0,8149	0,8276	0,8280	—	—
24 × 3	0	0,5	Δ_1	—	—	0,2342	0,0938	0,0162	0,0005	—
			ρ_{max}	—	—	0,9628	0,8043	0,8402	0,8276	—
			ρ_{min}	—	—	0,6866	0,7592	0,8140	0,8268	—
48 × 3	0	0,5	Δ_1	—	—	0,4853	0,3774	0,2342	0,0940	—
			ρ_{max}	—	—	0,9999	0,9955	0,9631	0,8356	—
			ρ_{min}	—	—	0,5928	0,6313	0,6366	0,7596	—
6 × 3	0,05	0,45	Δ_1	0,2360	0,1150	0,0036	0,0000	—	—	—
			Δ_2	0,3118	0,1574	0,0050	0,0000	—	—	—
12 × 3	0,05	0,45	Δ_1	0,3872	0,2619	0,0631	0,0010	0,0000	—	—
			Δ_2	0,4603	0,3874	0,0834	0,0008	0,0000	—	—
24 × 3	0,25	0,45	Δ_1	0,4841	0,4216	0,2418	0,0604	0,0070	0,0013	
			Δ_2	0,5317	0,4209	0,2626	0,0733	0,0012	0,0006	0,0002
6 × 3	0,25	0,25	Δ_1	0,0089	0,0037	0,0006	0,0001	0,0000	—	—
			Δ_2	0,0056	0,0008	0,0001	0,0000	0,0000	—	—
12 × 3	0,25	0,25	Δ_1	0,1045	0,0130	0,0117	0,0089	0,0052	0,0000	—
			Δ_2	0,1753	0,0017	0,0009	0,0006	0,0004	0,0000	—
24 × 3	0,25	0,25	Δ_1	0,3195	0,1004	0,0468	0,0474	0,0478	0,0469	0,0411
			Δ_2	0,3331	0,0441	0,0017	0,0016	0,0013	0,0010	0,0003
6 × 3	0,45	0,05	Δ_1	0,0057	0,0049	0,0047	0,0030	0,0012	—	—
			Δ_2	0,0003	0,0002	0,0001	0,0001	0,0000	—	—
12 × 3	0,45	0,05	Δ_1	0,0459	0,0460	0,0477	0,0486	0,0478	—	—
			Δ_2	0,0047	0,0007	0,0007	0,0007	0,0005	—	—
24 × 3	0,45	0,05	Δ_1	0,1550	0,0746	0,0734	0,0722	0,0729	—	—
			Δ_2	0,3284	0,0007	0,0007	0,0007	0,0005	—	—

the function of equation (5) circularization proceeds faster than for other functions, and as was stated it proceeds especially fast when $\sigma_1 = 0$. But even in this case it is fairly slow: when $\sigma_1 = 0$ the time required for circularization increases as n^2.

In tables 2–4 are shown the results of modeling for the function in equation (8) with an equilateral triangle as the initial configuration. Table 4 shows the dependence of the time required for circularization on the geometrical form of the initial configuration. It is greatest for an elongated rectangle. In this case circularization was not evident during the time used for modeling. In tables 2 and 3 the effect of parameters σ_1 and σ_2 on the circularization of a triangle is examined. It can be seen that an increase in σ_1 speeds up circularization at the beginning and slows it down at the end. When $\sigma_1 \neq 0$, circularization proceeds very slowly: with increase in n, the time required for circularization increases faster than n^4.* An analogous situation apparently results from an increase in d.

Table 2. Table of values of Δ_1, $\rho_{\max} = \max_i|A_i - A_{i+1}|$, $\rho_{\min} = \min_i|A_i - A_{i+1}|$ for f in equation (4) when $\gamma = \frac{1}{3}$ and $d = \frac{4}{3}$.

n	σ_1	σ_1	t	100	200	400	800	2000	4000	8000	11000
8×3	0	0,5	Δ_1	0,0803	0,0177	0,0058	0,0009	—	—	—	—
			$\rho_{\max}$	0,7552	0,7510	0,7505	0,7504	—	—	—	—
			$\rho_{\min}$	0,7460	0,7460	0,7502	0,7504	—	—	—	—
16×3	0	0,5	Δ_1	—	—	0,3110	0,2441	0,1950	0,1934	0,1930	0,1930
			$\rho_{\max}$	—	—	0,8072	0,7799	0,7229	0,6350	0,4903	0,3986
			$\rho_{\min}$	—	—	0,7985	0,7747	0,7206	0,6332	0,4889	0,3975

n	σ_1	σ_2	t	75	150	300	600	900	1200	3000	5550
12×3	0	0,5	Δ_1	0,3950	0,3161	0,2252	0,0554	0,0139	0,0052	—	—
			$\rho_{\max}$	0,8365	0,7955	0,7631	0,7491	0,7477	0,7575	—	—
			$\rho_{\min}$	0,7961	0,7758	0,7552	0,7457	0,7471	0,7473	—	—
24×3	0	0,5	Δ_1	0,5723	0,5382	0,4861	0,4502	0,3804	0,3515	0,2453	0,1921
			$\rho_{\max}$	0,9714	0,9262	0,8852	0,8605	0,8419	0,8325	0,8057	0,7840
			$\rho_{\min}$	0,8278	0,8398	0,8525	0,8484	0,8342	0,8268	0,8034	0,7829

* It has now been shown that in this case the time required for circularization increases as n^6.

Table 3. Table of values Δ_1 and Δ_2 for f in equation (4) when $\gamma = \frac{1}{3}$ and $d = \frac{4}{3}$.

n	σ_1	σ_2	t	75	150	300	450	600	750	900	1200
6×3	0,05	0,45	Δ_1	0,0070	0,0011	0,0004	0,0001	0	—	—	—
			Δ_2	0,0066	0,0001	0	0	0	—	—	—
12×3	0,05	0,45	Δ_1	0,1400	0,0346	0,0251	0,0185	0,0137	0,0103	0,0077	0,0044
			Δ_2	0,1157	0,0019	0,0010	0,0007	0,0005	0,0004	0,0003	0,0002
24×3	0,05	0,45	Δ_1	0,3301	0,1442	0,0263	0,0264	0,0137	0,0103	0,0077	0,0044
			Δ_2	0,3172	0,1558	0,0020	0,0010	0,0005	0,0004	0,0003	0,0002
6×3	0,25	0,25	Δ_1	0,0068	0,0034	0,0016	0,0015	0,0014	0,0008	—	—
			Δ_2	0,0003	0,0001	0	0	0	0	—	—
12×3	0,25	0,25	Δ_1	0,0224	0,0223	0,0214	0,0205	0,0193	0,0183	0,0174	0,0163
			Δ_2	0,0002	0,0001	0,0001	0,0001	0,0001	0,0001	0,0005	0,0005
24×3	0,25	0,25	Δ_1	0,0568	0,0573	0,0575	0,0578	0,0581	0,0584	0,0587	0,0590
			Δ_2	0,0002	0,0006	0,0006	0,0006	0,0006	0,0006	0,0006	0,0006
6×3	0,45	0,05	Δ_1	0,0071	0,0062	0,0046	0,0034	—	—	—	—
			Δ_2	0,0006	0,0005	0,0004	0,0003	—	—	—	—
12×3	0,45	0,05	Δ_1	0,0493	0,0507	0,0525	0,0533	—	—	—	—
			Δ_2	0,0001	0,0009	0,0008	0,0007	—	—	—	—
24×3	0,45	0,05	Δ_1	0,0759	0,0747	0,0739	0,0736	—	—	—	—
			Δ_2	0,0007	0,0004	0,0003	0,0003	—	—	—	—

In table 5 are shown the results of modeling for function f in equation (3). From table 5 (see also fig. 2) it is evident that with the increase of n, the time required for circularization increases as n^4.[†] Let us note also the interesting phenomenon of compression, which also takes place for the functions in equations (4) and (5) with $\sigma_1 \neq 0$ (on this subject see section 2).

It proved of interest to examine the case when

$$f = (1 - \sigma)r_0 + \sigma f_5.$$

Such a function is an example of a problem of the second type, when, in addition, $d = d(n) = 2R \sin (\pi/n)$. At first glance it might seem that if $\sigma > 0$ and is sufficiently small, there should be convergence to a regular polygon.

[†] This relationship has now been proved. An analogous result has been proved for function f in equation (4) with $\sigma_1 = 0$ and for function f in equation (5) with $\sigma_1 \neq 0$.

Table 4. Table of values $\Delta_1(t)$ and $\Delta_2(t)$ for f in equation (4) and for various d and initial conditions; $\gamma\frac{1}{3}$.

initial position	n	d	σ_1	σ_2	t	25	75	150	300	450	600	900	1200
triangle	12×3	$\frac{2}{3}$	0,25	0,25	Δ_1	0,0935	0,1233	0,2022	0,1913	—	—	—	—
					Δ_2	0,2601	0,0723	0,0057	0,0016	—	—	—	—
»	12×3	1	0,25	0,25	Δ_1	0,2382	0,1903	0,1914	0,1757	0,1505	—	—	—
					Δ_2	0,0038	0,0024	0,0012	0,0010	0,0010	—	—	—
»	12×3	3	0,25	0,25	Δ_1	0,1256	0,0447	0,0448	—	—	--	—	—
					Δ_2	0,1652	0,0019	0,0010	—	—	—	—	—
rectangle	$(3+15) \times 2$	0,5	0,25	0,25	Δ_1	0,4492	0,4743	0,5537	0,6536	0,6788	0,6702	0,6110	0,5721
					Δ_2	0,4784	0,3039	0,1375	0,0252	0,0043	0,0026	0,0017	0,0008
»	$(3+15) \times 2$	1	0,25	0,25	Δ_1	0,6474	0,5927	0,6064	0,5975	0,5853	0,5721	0,5529	0,5309
					Δ_2	0,0088	0,0043	0,0024	0,0017	0,0014	0,0013	0,0012	0,0012
»	$(3+15) \times 2$	2	0,25	0,25	Δ_1	0,1077	0,1073	0,1087	0,1103	0,1100	0,1096	0,1093	—
					Δ_2	0,3232	0,0004	0,0004	0,0004	0,0003	0,0003	0,0003	—
»	$(3+15) \times 2$	1	0,50	0,45	Δ_1	—	0,6252	0,5512	0,5054	0,4757	0,4501	0,4042	—
					Δ_2	—	0,0182	0,0093	0,0058	0,0043	0,0047	0,0032	—
»	$(3+15) \times 2$	1	0,45	0,05	Δ_1	—	0,6290	0,6307	0,6381	—	—	—	—
					Δ_2	—	0,0005	0,0003	0,0004	—	—	—	—
clover leaf (fig. 1)	27	$\frac{4}{3}$	0,25	0,25	Δ_1	0,4681	0,1070	0,0973	0,0753	0,0723	0,0520	0,0461	—
					Δ_2	0,9786	0,0018	0,0013	0,0013	0,0010	0,0007	0,0005	—

Table 5. Table of values $\Delta_3 = \Delta_1(t)$ for f in equation (3) when $\sigma = \frac{1}{2}$; initial configuration is on a triangle.

n								
6×3	t	5	9	18	25	50	75	—
	$\Delta_1(t)$	0,55	0,48	0,39	0,24	0,10	0,04	—
12×3	t	75	150	300	450	600	750	900
	$\Delta_1(t)$	0,56	0,47	0,28	0,16	0,11	0,08	0,05
24×3	t	1200	2400	4800	7200	9600	12000	—
	$\Delta_1(t)$	0,57	0,49	0,31	0,18	0,12	0,07	—

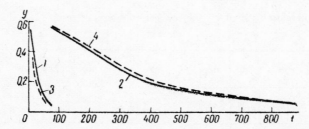

Fig. 2. A graph of the function $\Delta_1(t)$ and $\Delta_1(16t)$ for f in equation (7) when $\sigma = 1$ and the initial configuration is on an equilateral triangle. ———: $y(t) = \Delta_1(t)$ for $n = 6 \times 3$ (curve 1) and $n = 12 \times 3$ (curve 2); −−−−−: $y(t) = \Delta_1(16t)$ for $n = 12 \times 3$ (curve 3) and $n = 24 \times 3$ (curve 4).

This proved not to be so. To convince oneself of the unreliability of this assumption, it is sufficient to examine a regular polygon as the initial configuration. It is not difficult to see that if $\sigma > 0$ and the initial length of a side is greater than $2R \sin (\pi/n)$, the polygon transforms into a regular polygon with a yet greater length of side, so that with time the polygon becomes larger and larger. If to the contrary the initial length of the side is smaller than $2R \sin (\pi/n)$, this length will tend to zero as time tends to infinity. This function is an example of rules that from the local point of view seem good but from the global point of view cannot solve the problem. It is possible to prove that for such a rule when any $\sigma \neq 0$ and for all initial configurations, there is no convergence to a regular polygon.

2. STUDY OF THE LINEAR METHOD

The question of convergence to a regular polygon for an arbitrary initial configuration appears to be by no means a simple one, and in virtually no case have we succeeded in solving it. However, it is possible to consider the question of convergence if we start from initial configurations not far from the final one of a regular polygon. In such a case the problem can be substituted by a linear one, which of course is much easier to solve.*

We shall regard the position of a broken line with $r_1, \ldots, r_n$ apices as points in $2n$ dimensional phase space R^{2n}. The rules of motion for apices of the broken line are defined by a certain operator T in this space. The fixed points of this operator are essentially the stationary state of the broken line. Let r be some fixed point of the operator T. Set

$$A\rho = \lim_{\varepsilon \to 0} \frac{T(r + \varepsilon\rho) - r}{\varepsilon},$$

---

* We may remark that this option applies not only to the problem of circularization but also to the problem of straightening a line (as well as to other problems). In particular, it has proved possible to solve the straightening problem for the rules defined by equations (5) and (6) in reference 1.

where ρ is an arbitrary vector from R^{2n}. (We assume that this limit exists for any ρ.) Clearly, the operator A is linear, so it is natural to call this operator the linear part of the operator T.

It is obvious that for the convergence of the sequence $T^k r_0$, where r_0 is close enough to the stationary state, it is necessary that all eigenvalues of matrix A in their modulus not exceed 1, and sufficient that they all be of modulus strictly less than 1. Nevertheless, in our case matrix A always has eigenvalues equal to 1. This condition is related to the fact that fixed points (the stationary points of the broken line) of the operator are not isolated. Therefore a natural constraint on operator A, ensuring convergence in the vicinity of fixed point of the operator T, is that all eigenvalues of matrix A— with the exception of those that correspond to displacements along the manifold of fixed points of the operator (and these values, as has been said, are equal to 1)—be of a modulus strictly less than 1. Also, the rate of convergence is exponential: the deviation from the stationary state in time t is proportional to $|\lambda_{max}|^t$, where λ_{max} is the maximal eigenvalue of a modulus different from 1. For the problem of circularization of the second type, the fixed points of the operator form a manifold of three dimensions, and therefore there are three eigenvalues that are necessarily equal to 1. For the problem of circularization of the first type, the fixed points form a manifold (more strictly, a subspace) of four dimensions; thus there are four eigenvalues necessarily equal to 1.

On the basis of the above, the question of convergence for initial configurations close to the final one should be put as follows. First it is necessary to enumerate the linear part of the operator T, that is, operator A. This operator has eigenvalues (three or four) necessarily equal to 1. Further it is necessary to enumerate the remaining eigenvalues. If all of them are, according to their modulus, strictly less than 1, then for initial configurations close to the stationary state of the broken line, there is convergence to the stationary state; for the contrary case there is no convergence. The value of $|\lambda_{max}|$ characterizes the rate of convergence.

The discussion below will concern the rules that are defined by function f in equation (6). Because this function is linear, the operator it defines, acting in $2n$ dimensional space R^{2n}, is linear, and the considerations discussed above are applicable to the investigation of convergence from any initial configurations of the broken line. Thus we have to enumerate the eigenvalues of the operator and to elucidate the values of σ for which all of them, with the exception of four that are necessarily equal to 1, are of modulus less than 1. This will be done in this section. Let us remark that the considerations discussed below are useful for study of convergence from near the stationary state for any arbitrary function f, although the calculations become significantly more complex.

Let us first note that the center of gravity of the broken line, on the basis of equation (7), remains in place—specifically that the vector

$$\frac{1}{n} \sum_{i=1}^{n} r_1$$

does not change with time.

As was noted in section 1, regular broken lines of order 1 transform into ones similar to themselves. Because the center of gravity of the broken line remains in place, the set L_s of regular broken lines of order s with a fixed center (this set is a two-dimensional subspace of R^{2n}) is invariant with respect to T. It is easy to see that subspace L_s is characterized by vectors of the form $r = [\rho, \Phi^s \rho, \ldots, \Phi^{s(n-1)} \rho]$, where ρ is an arbitrary vector of the plane, and Φ is the transformation produced by turning clockwise through an angle $2\pi/n$. The two-dimensional subspaces L_s $(S = 1, \ldots, n-1, n)$ are orthogonal to one another with respect to the following natural scalar product in R^{2n}.

$$(r, r^1) = \sum_{i=1}^{n} (r_i, r_i'),$$

where $r = (r_1, \ldots, r_n)$ and $r' = (r_1', \ldots, r_n')$. When $s = 1$ the vectors from L_s are invariant with respect to T.

To find the eigenvalues of operator T, it is necessary to examine how the vectors from L_s transform. Let

$$r = (\rho, \Phi^s \rho, \ldots, \Phi^{s(n-1)} \rho) \in L_s.$$

We have [see equation (9)]

$$Tr = T(\rho, \Phi^s \rho, \ldots, \Phi^{s(n-1)} \rho) = T\{\Phi^{sj} \rho\} = \left\{ (1 - \sigma)\Phi^{sj} \rho \right.$$

$$\left. + \sigma \left[\tfrac{1}{2}(\Phi^{s(j-1)} \rho + \Phi^{s(j+1)} \rho) + \tfrac{1}{2} \operatorname{tg} \frac{\pi}{2} I(\Phi^{s(j+1)} \rho - \Phi^{s(j-1)} \rho) \right] \right\}$$

$$= \lambda_s \{\Phi^{sj} \rho\} = \lambda_s r,$$

where

$$\lambda_s = 1 - \sigma + \sigma \left(\cos \frac{2\pi s}{n} + \operatorname{tg} \frac{\pi}{n} \sin \frac{2\pi s}{n} \right)$$

$$= 1 - \sigma + \sigma \cos \frac{\pi}{n}(2s - 1) \Big/ \cos \frac{\pi}{n}. \tag{10}$$

Thus vectors from L_s are eigenvalues, and the eigenvalues of λ_s corresponding to them are two valued. Thus all eigenvalues $\lambda_1, \ldots, \lambda_{n-1}, \lambda_n$ are defined. As follows from equation (10), λ_1 and λ_n are equal to 1.

For convergence to the stationary state it is necessary that all eigenvalues be of modulus less than 1. From equation (10) it is evident that this is indeed so if $0 < \sigma < 1$ when n is even, and if $0 < \sigma < 1 - [\text{tg}\,(\pi/2n)]^2$ when n is odd; that is, for it to be asymptotic as $n \to \infty$ it is necessary that $0 < \sigma < 1$.

From equation (10) it is also possible to find λ_{max}, which determines the rate of convergence. It is evident that if only $0 < \sigma \leqslant [1 + \text{tg}\,(\pi/n)\sin(2\pi/n)]^{-1}$ when n is even and $0 < \sigma < \{1 + \text{tg}\,(\pi/n)[\sin(\pi/n) + s(2\pi/n)]\}^{-1}$ when n is odd, then $\lambda_{max} = \lambda_2 = 1 - \sigma_2 \text{tg}\,(\pi/n)\sin(2\pi/n)$. From this we see that if $0 < \sigma < 1$, then $1 - \lambda_{max} \sim 4\pi^2\sigma n^{-2}$ when $n \to \infty$. Therefore as n increases, the time required for circularization increases as n^2. This result is in conformity with computer modeling (see table 1).

Let $|\lambda_s| \leqslant 1$ for all s. We shall prove that in this case the phenomenon of compression of the broken line appears. To determine this we examine how the quantity $r^2 = \sum_{i=1}^{n} r_i^2$ transforms. Let $r = \sum_{s=1}^{n} r^{(s)}, r^{(s)} \in L_s$. Then $Tr = \sum_{s=1}^{n} \lambda_s r^{(s)}$. Because the subspaces L_s are orthogonal to one another, $r^2 = \sum_{s=1}^{n}(r^{(s)}, r^{(s)}), (Tr)^2 = \sum_{s=1}^{n} \lambda_s^2(r^{(s)}, r^{(s)})$. Because $|\lambda_s| \leqslant 1$ for all s, the quantity $r^2 = \sum_{i=1}^{n} r_i^2$ will not increase. For the same reasons the quantity $\sum_{i=1}^{n}(r_i - c)^2$ will also not increase if c is any vector of the plane. But it was shown that the center of gravity of the broken line remains in place, that is, the vector $1/n \sum_{j=1}^{n} r_j$ does not change. For this reason the quantity $\sum_{i=1}^{n}(r_i - \sum_{j=1}^{n} r_j)^2$ will not increase, thus signifying the compression of the broken line. This phenomenon, as remarked in section 1, was also studied by computer modeling (see table 1).

REFERENCES

1. Leontovich, A. M., Pyatetskii-Shapiro, I. I., and Stavskaya, O. N. (1970) Some mathematical problems related to morphogenesis. *Avtomatika i Telemekhanika* 4:94–107.

3. A Model of Inversion in *Volvox*

E. B. Vul and I. I. Pyatetskii-Shapiro*

The movement of a line in a plane that models the inversion of *Volvox* is studied. The basis of the model is the assumption that each individual cell is guided in its movements by knowledge of the geometrical positions of its nearest neighbors.

1. INTRODUCTION

The embryological development of both lower and higher organisms always includes complex movements of cellular tissues. It is the rule that a component of these movements is a folding of a layer, two examples of which are gastrulation and formation of the neural tube.

It is natural to suppose that the movements of cellular layers basically occur because of movements of individual cells and that each cell is guided in its movements by "knowledge" of the geometrical positions of its nearest neighbors and, perhaps, also by knowledge of a small number of global characteristics. In favor of such an assumption is the fact that as a rule an isolated part of a tissue can perform movements that are analogous to those it performs when it is part of the injured intact organism.[1]

Belousov has pointed out to the authors that from the geometrical point of view the inversion of *Volvox* may be the simplest and yet a very characteristic example. Let us recite some well-known facts pertaining to the development of *Volvox*.

Volvox is a green sphere whose interior contains a jelly-like substance and whose surface is a cellular monolayer. Each cell has flagella, whose coordinated movements propel *Volvox* through the water. The sphere has an opening called a phialopore. Several daughter individuals are usually found in the interior of a mature *Volvox* (fig. 1).

In 1919 the Russian biologist Kushakevich studied the embryology of *Volvox*. At an early stage of development the flagella of *Volvox* point inward. When the period of cell division ends, *Volvox* inverts and the flagella now point outward.

This paper reports studies of the movements of lines in a plane that model the inversion of *Volvox*. It is of course true that real inversion takes place

* First published in *Problemi Peredachi Informatsii* (*Problems of Information Transfer*) (1971) 7, no. 4:91–96.

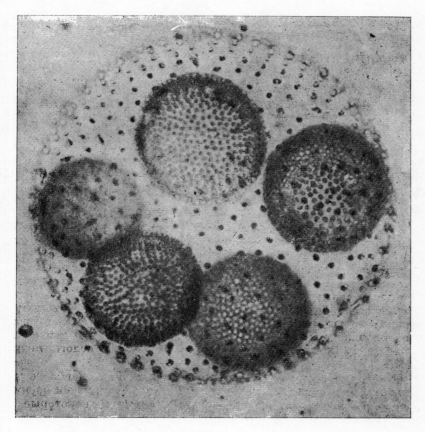

Fig. 1.

in three-dimensional space. Nevertheless, modeling of the analogous process in a plane has already led to interesting and sapid mathematical problems.

We may note that the modeling of movements of lines in a plane has been studied previously.[2,3] There the movements leading to straightening of lines and circularization were examined. The algorithms used in this paper are closely related to those studied in reference 2.

2. THE ALGORITHM OF MOVEMENT AND RESULTS OF MODELING

We model the tissue movements of *Volvox* as the movement of an open line in a plane. To each apex of the line is assigned a vector perpendicular to the chord joining adjacent apices. This vector will be referred to as a flagellum, and the apex itself as a cell. The cells are numbered in the natural order. The first and last are called terminal, the remaining ones internal.

In the model examined here time will be regarded as discrete. The position of each cell at the next moment of time depends on the position of neighbors at the previous moment, so that it is not important from which point movement begins.

We now give the motility rules. Let z_i be an internal cell that is not a neighbor of a terminal one, and let a_{i-1} and z_{i+1} be its neighbor. The distance between z_{i-1} and z_i is d_i. We connect points z_{i-1} and z_{i+1} by a segment of a line, and in the middle of the segment we erect a perpendicular in the same direction as the flagellum. We extend the perpendicular until it intersects the arc of a circle passing through z_{i-1} and z_{i+1} so that the angle corresponding to this arc is equal to $\frac{1}{3}$ of the sum of the angles at the apices z_{i-1} and z_{i+1}. The angle taken at each apex includes the angle of the flagellum directed to that apex (fig. 2). The point so obtained is designated z. The new positions of

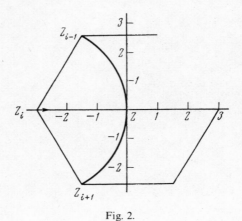

Fig. 2.

cells z_{new} are given by the equation

$$z_{new} = \alpha z + (1 - \alpha)z_i, \tag{1}$$

where

$$\alpha = \min\left[\tfrac{1}{2}, \max\left(\frac{\rho}{2}\right), 0\right], \quad \text{and} \quad \rho = \frac{\max(d_i, d_{i+1}) - cd}{|z_i z|}, *$$

where c is a constant whose value can be selected, usually taken to be 0.5–0.6; d is the distance between apices of a regular n-sided polygon. If $|z_i z| = 0$, z_{new} was taken to be equal to z_i.

We give the motility rules for terminal cells. Let z_1 be a terminal point. Then a line is drawn to pass through the two points that follow it. A ray is

---

* $|z_i z|$ is the distance between the points z_i and z.

drawn from the point neighboring the terminal point to the side opposite that of the flagellum at an angle ϕ_0 to the line constructed; on the ray a segment of length d is drawn, as shown in figure 3. The point so constructed is designated z. In our models the angle was usually taken to be equal to $2\pi/n$. The new position of the terminal point is described by equation (1).

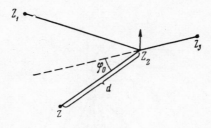

Fig. 3.

The point following the terminal one is constructed according to the same rule as an internal one, but here the angle is equal to half the sum of the angles at the neighboring internal point and at the point itself (fig. 4). From equation (1) it is evident that the coefficient at the old point is never less than $\frac{1}{2}$. This outcome is related to the divergence of the method, in which the position at the next moment of time is defined by the rule $z_{new} = z$, that is, $\alpha = 1$ in equation (1) and in which z is as defined in figure 3. The divergence is shown effectively in figure 5.

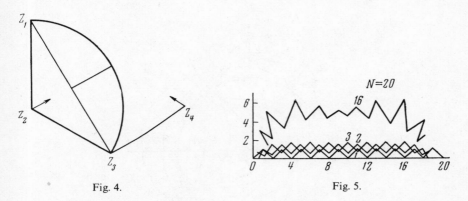

Fig. 4.

Fig. 5.

As initial configurations we have taken (1) an open regular polygon with flagella directed inward, (2) a straight line, (3) an angle, and (4) a periodic broken line. From the point of view of comparison with biological models, the most important is configuration (1) because movements starting from this configuration model the inversion of *Volvox*. Movements starting with configuration (2) model the well-known embryonic process of neural tube formation.

The process of inversion is shown in figures 6–8. We note that the time required for inversion rises rather rapidly with an increase in the number of cells. Thus, with $N = 6$, $T \approx 500$; $N = 15$, $T = 1000$; $N = 20$, $T = 3000$.

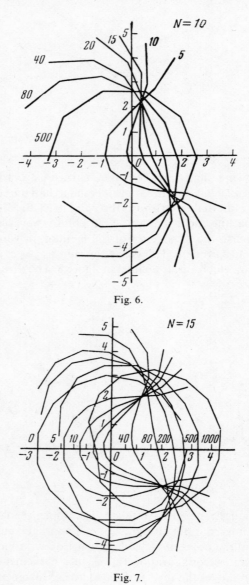

Fig. 6.

Fig. 7.

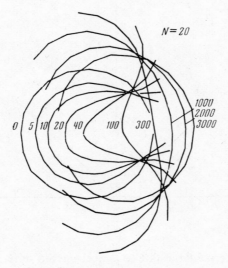

Fig. 8.

Figure 8 clearly shows that the overall time for inversion consists of two unequal parts, a time T_1 required for turning out followed by a time for turning back into a spherical form. For $N = 5$, $T_1 = 15$; $N = 15$, $T_1 = 40$; $n = 20$, $T_1 = 100$. Thus the greater part of the time is required for turning back into a spherical form. The long duration of inversion is due to the fact that after turning out, our line takes on a form that is close to a straight line. This development is confirmed by the following results of modeling. If an angle is taken as the initial configuration, closure into a spherical form takes place more rapidly than if the initial configuration is a straight line (figs. 9 and 10). This behavior suggested that the process would be accelerated

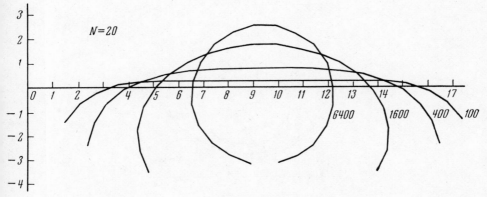

Fig. 9.

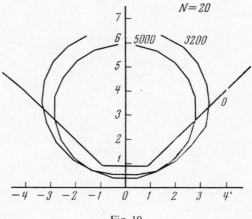

Fig. 10.

by changing an intermediate state to a more curved state. We did this in the following manner. At some intermediate state we improved the oval by making it a more angular figure (fig. 11). This alteration led to a significant acceleration of the process (fig. 12). We note that these dynamics of the process of inversion correspond well with experimental findings on the inversion process in one species of *Volvox*.[4]

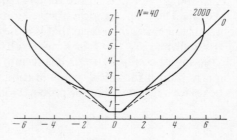

Fig. 11.

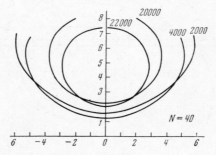

Fig. 12.

If a periodic broken line is taken as the initial configuration, the figure takes on the form of an elongated oval as a result of turning out, and to continue successfully, a correction is again necessary (fig. 13). We remark that in all calculated cases, inversion occurs notably faster if the distance taken for the construction of terminal points on the ray was greater than the distance of points in the initial configuration.*

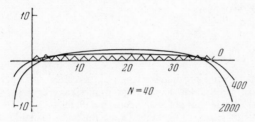

Fig. 13.

We thank L. V. Belousov, who as mentioned above originated the idea of modeling *Volvox*, for repeated discussions relating to the comparison of the present work with real biological data. The authors are grateful to A. M. Leontovich for many discussions of the present work. The authors express their sincere recognition for the great help of N. E. Vasilieva with computations. It should be mentioned that some of our results were presented at the International Embryological Conference held in Moscow in August 1969 and were discussed with Professors L. Wolpert and A. S. G. Curtis.

REFERENCES

1. Zakhvatkii, A. A. (1949) The comparative embryology of lower invertebrates. M. (Moscow ?) Sov. (or Current ?) *Sci.*
2. Leontovich, A. M., Pyatetskii-Shapiro, I. I., and Stavskaya, O. N. (1970) Some mathematical problems related to morphogenesis. *Avtomatika i Telemekhanika* 4: 94–107.
3. Leontovich, A. M., Pyatetskii-Shapiro, I. I., and Stavskaya, O. N. (1971) The problem of circularization in a mathematical model of morphogenesis. *Avtomatika i Telemekhanika* 2: 100–10.
4. Kelland, J. L. (1964) Inversion in *Volvox*. University Microfilms. Ann Arbor, Mich.

* In a paper to be published in *Problemi Peredachi Informatsii* we consider a model in which the distance between the terminal cells is less than 0.6 of the diameter of the circle at the initial state. Hence there is no intermediate position similar to a straight line. This accelerates the entire process of inversion and makes the pattern similar to the real one.

4. Modeling of the Processes of Sorting Out, Invasion, and Aggregation of Cells

A. B. Vasiliev, I. I. Pyatetskii-Shapiro,
and Y. B. Radvogin*

Local mechanisms that produce sorting out, invasion, and aggregation of cells are studied. It is shown that for the sorting out of cell aggregates whose dimensions are comparable to those that actually occur, it suffices to have 16–24 interactions per cell. It is further demonstrated that in the process of sorting out, the average cell moves a distance of 3–5 cell diameters. For the modeling of invasion it is shown that each cell has to interact with 2–3 layers of neighbors.

INTRODUCTION

A typical experiment[1-3] demonstrating the existence of sorting out is as follows. A suspension of disaggregated cells of two types derived from embryonic tissues is subjected to conditions (for example, rotation or shaking) such that the individual cells accidentally come into contact. As a result of these contacts, aggregates of cells appear that have a diameter of 0.1–1.0 mm and that contain 10^3–10^6 cells of both types, randomly interspersed. In each aggregate the process of sorting out, which starts after the stage of aggregation, results in the appearance of a small number of clusters, each of which consists of only one type of cell. In aggregates containing 10^3–10^4 cells, the number of components is two as a rule. The boundary of the aggregate is always composed of one type of cells.

The time required for sorting out is apparently only weakly dependent on the size of the aggregate and does not exceed 60 hours. According to the observations of many authors,[1,2,4] sorting out starts with the appearance of small clusters of cells that subsequently form the internal component. These initial clusters, which contain some tens or hundreds of cells, generally form within 20 hours. They grow at the expense of single cells of the internal type and usually begin to fuse within 40 hours after the beginning of aggregation. The final form of the internal component (or components) is as a rule far from spherical: it may consist of several smaller groups of cells joined together by bridges of cells. If there is one internal component, it is

* First published in Institute of Applied Mathematics of the Order of Lenin, Academy of Sciences of the USSR, Preprint 12 (1972).

not necessarily in the center. The minimal thickness of the outer component may be only one cell layer.

The behavior of individual cells during sorting has been studied by time-lapse cinematography and by photography at various intervals of time.[5,6] In both cases mixtures of heart cells and pigmented cells of the retina of the chick embryo were used. The authors, following the movements of individual pigmented cells in the aggregates, found that:

1. During sorting out, the individual pigmented cells move a distance of 30–50 μ and not more than 50 μ (the diameter of a pigmented cell is 7–10 μ).

2. Moving within the aggregate, pigmented cells change direction in an almost random manner; a cell moving toward another cell or aggregate of cells may reverse direction and moves as often toward an aggregate of pigmented cells as in the opposite direction.

3. Clusters of pigmented cells do not move.

A number of hypotheses have been proposed to account for this phenomenon. The most widely known is Steinberg's,[2] which consists of two parts: (1) the total energy of adhesion during sorting out increases, and the final state (when the cells are sorted out) is characterized by an energy of adhesion that is close to the maximum; (2) if the energy of adhesion between two types of cells i and j is designated W_{ij}, in order that the total energy of adhesion during sorting out should go to a maximum, it is necessary and sufficient that

$$0 < W_{11} < W_{12} < \frac{W_{11} + W_{22}}{2}. \tag{1}$$

The cells of the outer layer are type 1. Steinberg's hypothesis is that inequality (1) is an essential condition for sorting out to occur. Other hypotheses to explain the mechanism of sorting out have also been proposed.[3,4]

We have undertaken modeling of the process of sorting out and processes similar to it, that is, of cell invasion and aggregation, by assuming that movement of each cell is the result of forces of adhesion and therefore depends only on the distribution of other cells in direct contact with it. Of course, modeling on a computer cannot and should not aim to examine all local rules. We confine ourselves to the case where the behavior of the cell can be expressed in terms of the magnitude of its adhesion with each cell surrounding it, its capacity for active movement, and the geometrical pattern of the bonds it forms with other cells. The class of such rules of behavior includes, but does not reduce to the rules of, the hypothesis of Steinberg. Concurrent with our study, similar models were studied by Goel et al.[7] and Leith and Goel.[8]

Our modeling showed that (1) to obtain sorting out in aggregates, whose size is comparable to those actually observed, 16–24 bonds per cell suffice (see section 3). We also note that the dimensions of correct patterns of sorting out in the model are fully comparable with sections of a real cell aggregate.

Thus, an aggregate of 0.25-mm diameter has about 5×10^2 cells in diametric cross-section; about the same number is found in the aggregate of the model shown in figure 3.2b. (2) Sorting out occurs rapidly, and the average cell moves a distance not greater than 3–5 cell diameters (see section 3). It is of interest to compare this result with data on the path of a cell in real sorting out. Trinkaus and Lentz[5] observed, with the help of time-lapse cinematography, the movements of individual pigmented cells in an aggregate when intermixed with heart cells. They found that during sorting out (a time of the order of 50 hr), pigmented cells move through a distance not greater than 40 μ. which corresponds to 5–10 cell diameters. Counts by the computer give similar values, which, apparently, is an argument in favor of local rules of sorting out. (3) The quality of sorting out rises with an increase in active work by the motility mechanism of the cell (see section 4).

Some of the rules examined by us made possible the maximization of the total energy of adhesion and of sorting out and yet do not satisfy inequality (1) of Steinberg (see section 5). Thus it was found that to satisfy the first assumption of Steinberg the second is not essential, which is contrary to what Steinberg thought.

In addition to sorting out, some processes of morphogenesis related to it—the aggregation of cells into a mass of cells of different types and invasion by tumorous tissues—were also modeled (see section 6). Here we were particularly interested in the possibility of producing the phenomenon and studying the conditions essential for it.

1. DESCRIPTION OF THE MODEL

We shall assume that digits 0, 1, and 2 are located in each square of a flat grid. Here the zeros represent the extracellular space, and 1 and 2 the two types of cells that are sorting out. In the basic variant each cell of the grid has 8 neighboring cells distributed above it, above it to the right, to the right, and so forth (fig. 1.1a). There are 16 grid cells in the next layer, 24 in the third,

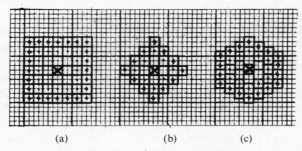

(a)   (b)   (c)

Fig. 1.1. Three types of grids used in modeling. □ cells at a distance of 1 from ⊠ ; ⊡ cells at a distance of 2 from ⊠ ; ⊞ cells at a distance of 3 from ⊠.

and so on. Two other variants, with 6 and with 4 neighboring grid cells, are shown in figures 1.1b and 1.1c. Any cell not farther than the third layer from a given cell is regarded as linked by bonds with it. Each bond has an energy of adhesion of W_{ij}^k, which depends on the type of cells i and j (the bond energy with 0 is taken to be zero) and on the distance k between them. Depending on the distance, k may have the values 1, 2 or 3. The energy of adhesion of a given cell is the sum of the energies of all the bonds.

Cells can move (change positions) according to the following rules. Initially a cell is chosen at random and is then regarded as active. Then one of the eight neighboring cells of opposite type is chosen, and the adhesion energy of the two cells chosen is computed. Let W_a be the energy of adhesion of the active cell and W_c that of the neighboring cell. For an exchange, the neighboring cell is chosen that gives on exchange the largest increase in $W_a + \alpha W_c$, where α is a parameter of the rules. The largest increase can also be negative. An exchange of the two cells takes place if the increase plus a certain quantity F_i, which depends on the type of active cell, is not less than zero.

We note the relation of this model to that studied by Goel et al.[7] and to another model of ours.[9] These two models are closely related to each other and are special cases of the model studied here (in reference 7 a slightly different organization of the work of the cell in time is taken). The basic difference is that cells form bonds with four cells in the model in reference 7 and with eight in the reference 9 model. Furthermore, the motility rules are simpler and can be obtained from those described above by setting $\alpha = 0$ in reference 7 and $\alpha = 0$ or $\alpha = 1$ in reference 9. With such a choice the translocation of a cell maximizes the increase of its energy of adhesion ($\alpha = 0$), or the sum of this energy and the energy of adhesion of the partner in the exchange ($\alpha = 1$). In addition, the model in reference 7 includes only two types of cells: 0 (the extracellular medium) and 1 (the cells themselves); thus the situation represents aggregation of one type of cells rather than sorting out.

In a recent paper Leith and Goel[8] expand their model by introducing certain changes. The fundamental ones are that a cell may translocate in one step to greater distances and that the direction of movement of a cell may be conserved for several translocations. It is shown that this method improves sorting out.

All these models simplify reality in one of the following basic ways:

1. The regular homogeneous structure of the bonds given by the rules of the model differs from reality, where every cell has its own chance pattern of contacts. The effect of this simplification can be evaluated to some degree by changing the number and geometrical structure of bonds in the model and by observing the effect that this produces. The number of bonds proved to be the only important factor (see section 2, page 50).

2. In the model the cells are distributed in a plane and not in real three-dimensional space. It would appear evident that it would be more difficult for cells to achieve sorting out by moving only in one plane within the cell mass. If this is so, the quantities describing the quality of sorting out (e.g., the relative magnitude of the boundary between different types of components) should be poorer than in three dimensions for the same diameter of the aggregate, and they can be regarded as a lower limit for what would occur in three dimensions. Nevertheless, the basic qualitative effects in two- and three-dimensional models are apparently the same. Goel et al.,[7] in addition to studies in two dimensions, investigated a three-dimensional case (a cubic grid) where each cell has 26 bonds and found that the degree of aggregation increases compared to the two-dimensional case, but not by much. Sections of the three-dimensional structure, in the words of the authors, closely resemble the two-dimensional pattern.

2. CONDITIONS FOR SORTING OUT

We present without proof a finding of one of the authors (Vasiliev) that makes it possible to relate the rules of the behavior of cells in the model and the configurations of cells toward which these rules lead, and also to formulate the conditions that lead to sorting out.

In the rules of cell motility (section 1) there occur the quantities W_a, the energy of adhesion of the active cell, and W_c, the magnitude of the adhesion energy of its neighbors of the other type. These quantities, and therefore the behavior of the active cell, are determined by the active cell's immediate neighborhood. It appears that in many cases the behavior can be described in terms of maximizing a quantity that depends on the configuration of all the cells. In particular, let one of two constraints hold:

1. The energy of a bond, W_{ij}^k, does not depend on its length k, because k does not exceed some values of R and is zero when $k > R$.

2. The quantity W_{ij}^k is arbitrary, but the parameter of the rules, α, is equal to one.

We evaluate the total adhesion energy of type-1 cells in a given configuration by writing

$$E_{11} = \sum_{k=1}^{R} W_{11}^k N_{11}^k,$$

where, depending on the configuration, N_{11}^k is the number of contacts of length k between type-1 cells.

Analogously, we evaluate E_{22} (E_{12}), the total energy of adhesion between type-2 cells (and correspondingly between type-1 and type-2 cells) and also the quantities

$$E^\alpha_{12} = E_{11} + E_{12} + \alpha E_{22} + (1 - \alpha)\frac{W_{12}}{W_{11}}E_{11}$$

$$E^\alpha_{10} = E_{11} + E_{12}$$

$$E^\alpha_{21} = E_{22} + E_{12} + \alpha E_{11} + (1 - \alpha)\frac{W_{12}}{W_{22}}E_{22}$$

$$E^\alpha_{20} = E_{22} + E_{12}.$$

The above constraints on the rules of behavior can be expressed this way: The active cell of type i "chooses" that of its neighboring cells of type $j, j \neq i$, with which an exchange of positions gives the greatest increase in the quantity E^a_{ij}. The exchange occurs if the greatest increase found when summed with F_i is not less than zero. Because in the usual situation an exchange of positions happens almost exclusively between cells of types 1 and 2, it is sufficient to evaluate only the quantities E^a_{12} and E^a_{21}. In this case it follows from the new formulation that, for example, the behavior of cells of type 1 is directed toward the maximization of the quantity E^a_{12}, which depends on the overall configurations of the cells. It is easy to convince oneself that when $\alpha < 1$, the quantities E^a_{12} and E^a_{21} in different cases depend on N^k_{11}, N^k_{12}, and N^k_{22} and, therefore, can reach a maximum in various configurations. In this case there are two competing processes in the cell mass: activation of 1 leads to an increase in E^a_{12} (also possible is a decrease in E^a_{21}), and the reverse. Each process can be slowed or stopped by a suitable choice of F_1 and F_2. Thus, choosing F_1 to be larger than the greatest possible value of the increase in E^a_{12}, we obtain a selective braking of cells of type 1 (because the conditions of exchange for them will not be fulfilled) and an increase in E^a_{21}.

When $\alpha = 1$, the expressions E^a_{12} and E^a_{21} coincide and give the total adhesion energy of the cell aggregate:

$$E = E_{11} + E_{21} + E_{22}.$$

That is, the behavior of all cells will be directed toward maximizing the quantity E, as is conjectured in the hypothesis of Steinberg.

It is possible to show that after sorting out has occurred, E^a_{21} is maximal when the inner component is of type 2 if

$$\alpha W^k_{11} \leqslant W^k_{12}$$

$$W^k_{22} + \alpha W^k_{11} - (1 - \alpha)W^k_{12} \geqslant 0, \tag{A}$$

but the lower expression must be positive when $k = 1$ and not increase with increase in k.

If $\alpha = 1$ (when $E^a_{21} = E^a_{12} = E$, Steinberg's case), relation (A) becomes the condition

$$W^k_{11} \leqslant W^k_{12} \leqslant \frac{W^k_{11} + W^k_{22}}{2}. \tag{B}$$

Generalizing the conditions of Steinberg to the case of cells having contacts of arbitrary length and given $\alpha = 0$,

$$W^k_{12} \leqslant W^k_{22}. \tag{C}$$

The conditions are analogous for maximizing E^k_{12} when the internal component is a type-1 cell. The corresponding inequalities are obtained from relation (A) by changing W^k_{11} into W^k_{22}:

$$\alpha W^k_{22} \leqslant W^k_{12}$$

$$W^k_{11} + W^k_{22} - (1 - \alpha)W^k_{12} \geqslant 0.$$

Condition (A) strongly depends on the value of α. For example, inequality (B) (corresponding to $\alpha = 1$) holds only for the nonspecific adhesion of cells when $W_{11} \leqslant W_{12} \leqslant W_{22}$. If $\alpha = 0$ [inequality (C)], the specific case $W_{11} > W_{12} < W_{22}$ is possible (but not $W_{11} < W_{12} > W_{22}$). Therefore it is important to clarify what value of this parameter most adequately corresponds to the real motility rules.

It should be stressed that conditions (A)–(C) are merely sufficient; that is, in principle there can be values of W^k_j that do not satisfy (A)–(C) but nevertheless give sorting out (see section 4).

We shall return to the question of when the quantity E^a_{ij} reaches a maximum in section 3. The computer experiments presented there show that the maximum is in fact reached if the number of cells in the aggregate is not excessively large. With a large number of cells the model gives a value of E^a_{ij}, which is near the maximum but not identical with it (for example, a sorting out that results in two internal components).

3. The Dependence of Sorting Out on Geometrical Parameters

In this section experiments are described that demonstrate the possibility of sorting out in the model and its dependence on the dimensions of the aggregate and on the percentage composition of the two cell types. Here the values of α, of the adhesion energy W^k_{ij}, and of the energy of the motility mechanism F_i have been chosen to obtain the best possible pattern of sorting out.

Two basic variants of the parameters are used.

A. $\alpha = 0$; $F_1 = -1000$; $F_2 = 1$; $W^k_{11} = W^k_{22} = 4 > W^k_{12} = 3$,

      $K = 1, \ldots, R$, $W^k_{ij} = 0$ when $k > R$; $R = 1, 2, 3$.

In this case a cell undergoing an exchange maximizes its own energy of adhesion, or it maximizes the total adhesion energy of cells of its own type. The parameter F_i is chosen so that a type-1 cell, having become active, would not complete an exchange with any configuration of neighboring cells; only cells of type 2 can increase their adhesion. It is possible to show that their energy of adhesion reaches a maximum in a configuration in which a central mass of type-2 cells is surrounded by a layer of type-1 cells and in which the boundary between type-1 and type-2 cells has the minimum length (see also section 2).

B. $\alpha = 1$; $F_1 = F_2 = 1$; $W^k_{11} = 3$, $W^k_{12} = 4$, $W^k_{22} = 6$,

$k = 1, \ldots, R$, $W^k_{ij} = 0$ when $k > R$; $R = 1, 2, 3$.

As has been noted in section 2, if $\alpha = 1$, the rules of cell motility maximize the total adhesion energy of the entire configuration. For the chosen W^k_{ij}, its maximum is reached in the same configuration as case A. The motility rules and values of adhesion of cells in case B correspond to the hypothesis of Steinberg.

Both variants give results close to each other, but B gives sorting out for the great majority of initial configurations. Other choices of parameters (not A or B) are shown in the legends to the figures.

As in the initial configuration in modeling, a portion of the grid is chosen (a square or an approximation to a circle) and is randomly filled with 1, 2, and 0. The remaining portion of the grid is filled with 0. By τ we shall signify the number of steps during which the average type-1 or type-2 cell becomes active once. At times $t = \tau, 2\tau, 3\tau \ldots$, the following characteristics are determined:

1. The fraction of aggregated i cells, $i = 1, 2$

$$A_i(t) = \frac{1}{\tau} \times \text{(number of } i \text{ cells, all neighbors of which are type } i\text{)}$$

2. The fraction of i cells that form the boundary of the aggregate

$$\Gamma_i(t) = \frac{1}{\tau} \times \text{(number of } i \text{ cells that have 0-neighbors)}$$

3. The fraction of i cells that have undergone exchange in time τ

$$\Delta_i(t) = \frac{\text{(number of } i \text{ cells that have completed exchanges during } \tau\text{)}}{\text{(total number of } i \text{ cells)}}.$$

The numbers $A_i(t)$ and $\Gamma_i(t)$ are calculated for the period of time $(t - 1)\tau - t\tau$ as follows. At each selection of an active cell, the program analyzes its type, its state of aggregation, and membership in a boundary, and 1 or 0 is added in the corresponding register; this is normalized at time τ. Because the selection of an active cell is made at random and with equal probability for all type-1 and -2 cells, the values of $A_i(t)$ and $\Gamma_i(t)$ are close to the true fractions of aggregated and boundary i cells.

In the initial distribution certain cells are tagged in order to observe their movements. The rules of tagging and the notation of cells in the figures are given below:

1 cell	1 or +
2 cell	2 or ×
tagged 1 cell	5 or $\stackrel{\smile}{+}$
tagged 2 cell	6 or $\stackrel{\smile}{\times}$
0 cell	0 or blank
tagged 0 cell	4 or .

In a given experiment it was assumed that sorting out had occurred if at the end of the experiment the boundary of the aggregate consisted of not less than 90 percent of cells of a single type:

$$\Gamma(t) = \frac{\{\max \Gamma_1(t), \Gamma_2(t)\}}{\Gamma_1(t) + \Gamma_2(t)} \geqslant 0.9$$

and if the overall fraction of aggregated cells, $A(t) = A_1(t) + A_2(t)$, was greater than at the start.

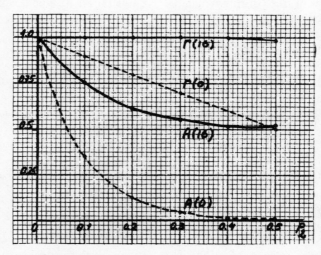

Fig. 3.1. The dependence of the fraction of aggregated cells and the composition of the boundary Γ on ρ_2 in the beginning and at the end of sorting. Time in τ; variant A, $R = 3$, $\rho_1 = 1 - \rho_2$, size of the aggregate 64×64.

Figure 3.1 shows the dependence of $A(t)$, $\Gamma(t)$ with $t = 15\tau$ and $A(0)$, $\Gamma(0)$ on ρ_2, the density of type-2 cells in the aggregate. In the given case sorting out occurs for all values $\rho_2 \leqslant 0.5$.

A pattern of sorting out typical for an aggregate of no great size is shown in figure 3.2b (fig. 3.2a shows the initial configuration). It can be seen that type-2 cells gather themselves into one internal aggregate surrounded by a surface layer of type-1 cells. The boundary between type-1 and -2 components is convex, and the thickness of the boundary layer of type-1 cells varies between 2 and 4 cells. This configuration has an energy of adhesion that is almost maximal (in the given case it is 2 % less). For brevity, we shall refer to a pattern of this type as the simplest.

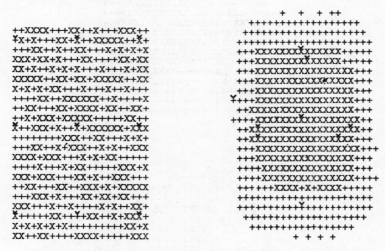

Fig. 3.2a. Initial state.   Fig. 3.2b. state at time 15τ.

Variant B, $R = 3$ (bonds of length 1 are absent, that is, $W^1_{ij} = 0$); $\rho_1 = \rho_2 = \frac{1}{2}$; size of the aggregate 22 × 22.

We examine the changes in sorting out when sizes increase and the dependence of such changes on ρ_1 and ρ_2. When $0.4 < \rho_2 < 0.6$ (a type-2 cell forms the internal component), an increase in dimensions leads to the patterns shown in figures 3.3a and 3.3b. The number of aggregates of the type-2 component increases, or the aggregates cease to be composed of one cell type, that is, they contain spaces filled with type-1 cells. As a rule the boundary is not convex. However, the adhesion energy of this pattern continues to be close to the maximum; in other words, the length of the boundary Γ_{12} between the aggregates of type-1 and type-2 cells is close to the minimum (fig. 3.4).

When $\rho_2 < 0.35$, the amount of the type-2 component increases with the increase in the surface of the aggregate, its form continues to be rounded,

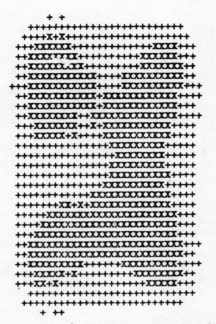

Fig. 3.3a. Variant B, $R = 3$, $\rho_1 = \rho_2 = \frac{1}{2}$; sizes of the aggregate 29×29; state at time $t = 15\tau$.

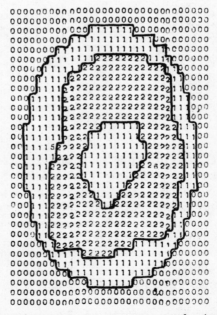

Fig. 3.3b. Parameters α, W_{ij}^1, and F_i as in variant B, but $W_{ij}^2 = \frac{1}{2}W_{ij}^1$, $W_{ij}^3 = \frac{1}{3}W_{ij}^1$, $R = 3$, $\rho_1 = \rho_2 = \frac{1}{2}$; diameter of the aggregate 27×27; state at time $t = 15\tau$.

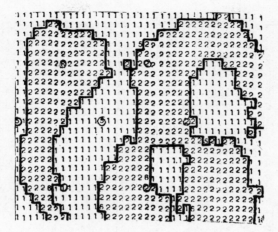

Fig. 3.3c. Variant A, $R = 3$, $\rho_1 = \rho_2 = \frac{1}{2}$; portion of the aggregate of size 64×64; state at time $t = 15\tau$.

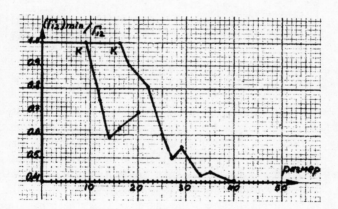

Fig. 3.4. The dependence of the relative length of boundary between type-1 and type-2 components on the dimensions of the aggregate with $R = 1$ (left curve) and $R = 2$ (right curve). Variant B, $\rho_1 = \rho_2 = \frac{1}{2}$.

and its surface varies around a value V, which is characteristic for a given selection of parameters (figs. 3.5a and 3.5b). When $\rho_2 > 0.7$, there is no sorting out in the strict sense of the word no matter what the dimensions of the aggregate, because the outer component of type-1 cells forms less than 90% of the boundary.

Experiments have shown that there is a certain critical size K: if the size of the aggregate is less than K, the pattern of sorting out is always the simplest. In figure 3.4 the critical size corresponds to the points of the first break in the curves. The magnitude of K depends on the relationship between ρ_1 and ρ_2

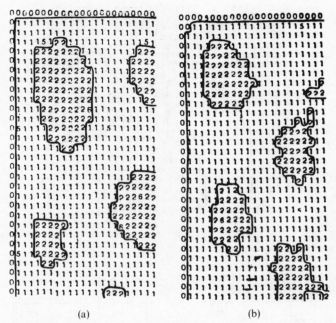

(a)   (b)

Fig. 3.5. Variant A, $R = 3$; $\rho_1 = 0.7$, $\rho_2 = 0.3$ (a) and $\rho_1 = 0.8$, $\rho_2 = 0.2$ (b); a portion of the aggregate of size 64×64; state at time $t = 15\tau$.

and decreases with decreasing values of ρ_2. The most frequently observed value of K is 24 (the shape of the aggregate is a square). We note that in all cases investigated, the character of the sorting-out pattern as sizes increase does not depend on W_{ij}^k, F_i, and α if these values permit sorting out at all; only the magnitudes of K and V change. Given a certain ρ_2, with the help of K it is possible to evaluate the quality of the motility rules of the cell. The higher the K, the larger the size of the aggregate that still shows good sorting.

The experiments described show that the model can produce sorting out if the concentration of cells in the internal component is not excessively large ($\rho_2 \leqslant 0.6$). The quality of sorting out decreases strongly with an increase in the size of the aggregate above a certain critical value. This value, which depends on the concentration of internal cells, falls as the concentration decreases.

This behavior of the model conforms with data on real sorting out, in which the number of internal continuous components increases with an increase in the size of the aggregate or a decrease in the fraction of cells of the internal component.[2,6]

1. Common to all experiments was the finding that sorting out proceeded rapidly. The curves in figures 3.6a and 3.6b show the mean number of exchanges completed by cells in the time τ for aggregates of various sizes and

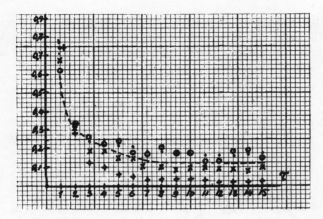

Fig. 3.6a. Mean number of exchanges completed by one cell in time τ. Variant B, $\rho_1 = \rho_2 = \frac{1}{2}$.
$\bigcirc$: aggregate 22×22, $R = 1$; $\times$: the same, $R = 2$; $+$: the same, $R = 3$; $\bullet$: aggregate 64×64, $R = 1$.

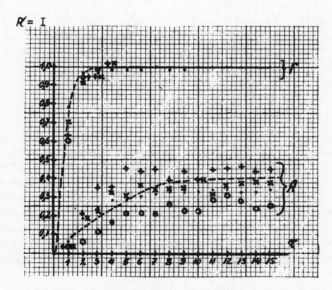

Fig. 3.6b. The increase of $A(t)$ and $\Gamma(t)$ for the same aggregates.

for various numbers of bonds. In figure 3.1 are given curves of the increase in the fraction of aggregated cells and in the fractions of units in the boundary (in the experiments the 2s aggregate in the interior). It can be seen that the characteristic pattern of sorting out stabilizes when $t = 5\tau$. One can take this value as the time required for sorting out. The mean number of exchanges completed by cells in time 5τ depends (other parameters remaining fixed) on the size of the aggregate and on the type of bonding, but it is always less than 3. Under the same conditions the number of exchanges in time 15τ is not greater than 6; the pattern of sorting out after 5τ shows almost no change (cf. figs. 3.7a and 3.7b).

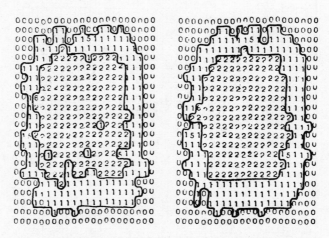

Fig. 3.7a. State at time $t = 5\tau$.   Fig. 3.7b. State at time $t = 10\tau$.

Fig. 3.7. Variant B, $R = 3$, $\rho_1 = \rho_2 = \frac{1}{2}$, size of the aggregate 18×18.

In all cases the distances the cells move from their starting positions do not exceed 5, but it is possible to observe some increase in these distances as the quality of sorting out improves. The cells of the internal component, which are initially on the periphery of the aggregate, have to take a longer path than cells that are initially in the interior. It is natural that the time for sorting does not depend on the size of the aggregate because given the local nature of the motility rules, distant parts of the grid behave independently.

2. The number of bonds formed by one cell has an influence on the behavior of the model. Thus the model of Goel et al.[7] and model B in reference 9 differ only in the number of bonds formed by one cell (4 and 8, respectively), yet the results obtained are significantly different: sorting out is observed only in the second model.

The experiments described below show that 16 contacts suffice to ensure sorting out in an aggregate whose size is comparable to real sizes. In the model the maximum number of bonds formed by one cell depends on the distance of its action and on the type of grid. We shall describe two series of computer experiments. In the first, a fixed grid is used and the maximal length of bonds varies, whereas in the second, both the length of bonds and the type of grid vary.

Series 1. The grid is square, and each cell has 8 cells at distance 1, 16 at distance 2, and 24 at distance 3. The set of bonds for a cell varies according to the number of all bonds of given length in accordance with equation $W_{ij}^k = 0$ for given k and all i and j. Because $k \leqslant 3$, seven types of bonding are possible (if one excludes the case of their nonexistence). As in section 1, for each type there is a critical value K and percentage of aggregated cells, A. The results and the number of bonds of each type are shown in table 3.1.

Table 3.1. The dependence of K and A on the type of bonding. Each value of K was obtained in one series of experiments; the value of A is given for aggregates of size 22×22. Asterisks indicate cases where the type-2 component has "holes" (see text). The values of α, F_i, ρ_1, and W_{ij}^k for nonzero bonds as in variant B, $\rho_1 = \rho_2 = \frac{1}{2}$.

Number of series	1	2	3	4	5	6*	7*
Lengths of nonzero bonds	1	1, 2	1, 2, 3	2, 3	1, 3	3	2
Number of bonds	8	24	48	40	32	24	16
K	10	17	20	23	19	...	...
A	0.25	0.42	0.45	0.48	0.44	0.32	0.19

It can be seen that the critical values K and A monotonically increase with increasing numbers of bonds and increase more rapidly with addition of bonds of length 2 to those of length 1 (columns 1 and 2). Inclusion of the next layer of bonds produces a smaller effect.

Let us note the extreme cases: $W_{ij}^k = 0$, $k > 1$, when each cell makes contact with 8 neighbors, and $W_{ij}^k > 0$, $k = 1, 2, 3$, when the cell makes contacts with its 48 nearest neighbors. The corresponding patterns of sorting out are essentially different. In the case of short bonds (fig. 3.8), the boundary of inner layer 2 is irregular, its length is far from the minimal, and there is more than one aggregate of type-2 cells. With long contacts the pattern of sorting out is the correct one (fig. 3.2b). In the first case the size of the aggregate is greater than the critical, in the second less than the critical, but the difference between the patterns of sorting out is observed for all large sizes.

It is worth observing that when the cell has only bonds of length 2 (or only of length 3, columns 6 and 7 of table 3.1), one observes an unusual final pattern, where the internal type-2 cell component has spaces filled with type-1 cells. An example of such a pattern, shown in figure 5.2, can be explained as follows. The configuration that is composed of alternating parallel layers of type-1 and type-2 cells, surrounded by a common boundary layer of 1s, is at a point of local maximum adhesion energy for this type of bonding.

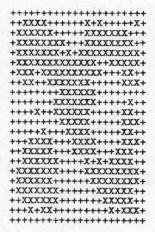

Fig. 3.8. Variant B, $R = 1$, $\rho_1 = \rho_2 = \frac{1}{2}$; size of the aggregate 22×22; state at time $t = 15\tau$. Compare with figure 3.2b (where $R = 3$).

Series 2. Grids of three types are used: a square with eight neighbors for each cell, a hexagonal one, and a square one with four neighbors (fig. 1.1). On each grid three types of bonding with maximal lengths of 1, 2, and 3 are modeled. To make the results comparable with those of Goel et al.,[7] we assumed the same initial configuration that appeared in their paper (fig. 3.9a), where the aggregate consists only of type-1 cells. The parameters

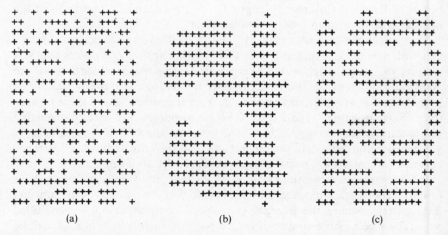

(a)   (b)   (c)

Fig. 3.9. (a)–(c). Comparison with the model of Goel, $\rho_1 = 0.58$, $\rho_2 = 0$, $\rho_0 = 0.42$; size of the aggregate 20×20.

Fig. 3.9a. Initial state. Fig. 3.9b and c. State at time $t = 15\tau$. Square grid, where each cell has 4 neighbors. (b) $R = 3$ and (c) $R = 1$.

W^k_{ij} and F_i in all nine experiments were the same as in the model of Goel et al. The maximum energy of adhesion is reached with a compact aggregation of type-1 cells having a minimal length of boundary between the 1 and 0 components.

In this series of experiments the quality of sorting out is evaluated according to its final pattern, where the number α of cells of type 1 having eight nearest neighbors of the same type is counted. The results are presented in table 3.2. In figures 3.9b, 3.9c, 3.10a, 3.10b, 3.11a, and 3.11b the best and worst patterns of sorting out obtained using each of the three types of grids are shown.

Table 3.2. Dependence of A on the lengths of bonds and on the type of grid. $\rho_1 = 0.58, \rho_2 = 0, \rho_0 = 0.42, \alpha = 0, W^k_{11} = \{^1_0, F_1 = +1.$ Size of the aggregate 20×20.

Lengths of nonzero bonds — Type of grid	1	1 and 2	1, 2, and 3
Square, 4 neighbors	0.10	0.34	0.45
Hexagonal, 6 neighbors	0.18	0.42	0.49
Square, 8 neighbors	0.26	0.43	0.62

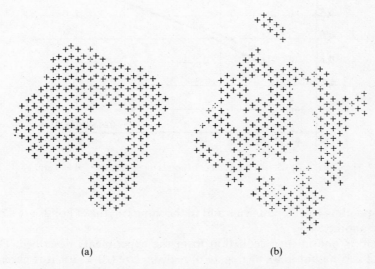

(a)   (b)

Fig. 3.10. Hexagonal grid, (a) $R = 3$ and (b) $R = 1$; state at time $t = 15\tau$.

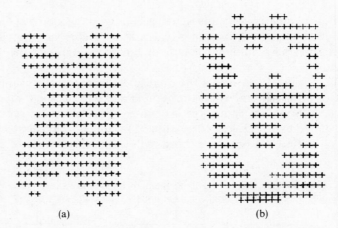

Fig. 3.11. Square grid, where each cell has 8 neighbors, (a) $R = 3$ and (b) $R = 1$; state at time $t = 15\tau$.

Table 3.2 and a comparison of the pattern of sorting out permit the conclusion that its quality is basically determined by the number of bonds formed by each cell and to a lesser degree by their maximal length. Figure 3.12 is a graph of the dependence of α on the number of contacts. The curves

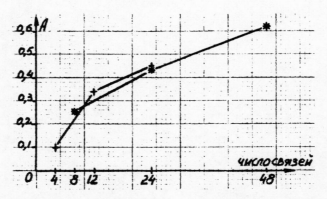

Fig. 3.12. The dependence of A on the number of bonds for square grids. $*$ = 8 grid, $+$ = 4 grid.

corresponding to different grids and to the same lengths of bonds are similar to one another.

There is one further deduction from the experiments described. For a correct sorting out in an aggregate of realistic size (20×20), it is necessary to have at least 16–24 contacts for each cell.

4. The Influence of Cell Motility

The experiments presented in this section concern the question of how cell motility and differences in motility among different cell types affect the behavior of the model. The answer to this question depends on the value of parameter α. If the selected active cell is i, it will complete an exchange with the neighboring cell type j that leads to the greatest increase in the quantity

$$\Delta_a(i, j) = \Delta W_i + a\Delta W_j,$$

where ΔW_i is the increase in the adhesion energy of the active cell and ΔW_j is the increase for its partner in exchange. Exchange will occur in the selected direction if $(\Delta a)_{max} + F_i \geqslant 0$, where F_i is the height of the potential barrier that can be surmounted by a type-i cell. The quantity F_i can be regarded as a measure of the strength of the cell's motility apparatus.

In section 2 it was stated that $\Delta_a(i, j)$ can be replaced in these rules by the increase, after an exchange, of the function E_{ij}^a, which depends on the entire configuration of cells. Thus, the movements of type-i cells are directed toward maximizing the value of E_{ij}^a. The form of E_{ij}^a (and the point of its maximum) is determined by the parameters W_{ij}^k. The values of F_1 and F_2 determine the rate of increase in E_{12}^a and E_{21}^a.

The influence of F_1 and F_2 on the behavior of the model was studied for the extreme values $\alpha = 0$ and $\alpha = 1$.

1. $\alpha = 1$. In this case the movement of cells is directed toward raising the total energy of adhesion, $E = E_{21}^1$ (see section 2), that is, the model attempts to reach a maximum E. The form of the configuration of the maximum is determined by the value of W_{ij}^k. Here they are chosen so that the maximum has the form of a sorting out. [W_{ij}^k satisfies conditions (B) in section 2.] Two series of experiments are described below.

In the first series, $F_1 = F_2 = F$ varies from -40 to $+1000$. The latter value is selected so that for any configuration and for any neighboring i and j cells, the inequality $W_i + W_j + 1000 > 0$ would hold. In this case any active cell must necessarily move if it has at least one neighbor cell of the opposite type. For the value -40 the conditions for exchange are not fulfilled for any cell, and the initial configuration will not change with time.

The results are shown in figure 4.1. It can be seen that the number of aggregated cells falls with the decrease of F from $+2$ to -6, and with $F = -40$ the number coincides with the number of aggregated cells in the initial configuration. With the rise in F from $+2$ to $+1000$, this number also falls but not significantly. Thus to obtain sorting out, it is essential that the cells have the ability to move even in those cases where such movement leads to a decrease in the total energy of adhesion. In the biological interpretation this condition corresponds to the active work of the motility apparatus of the cell. When $F < 0$ (i.e., when cells move only when there is an increase in E), sorting out becomes poorer or ceases entirely.

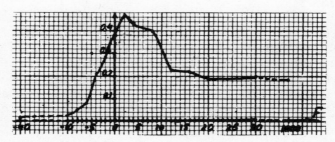

Fig. 4.1. The dependence of the fraction of aggregated cells on F ($F = F_1 = F_2$). Parameters α and W_{ij}^k as in variant B, $R = 3$, $\rho_1 = \rho_2 = \frac{1}{2}$; size of the aggregate 20×20; state at time $t = 15\tau$.

In another series the values of F_1 and F_2 were chosen in order to slow down cells of one type (the cell of the slowed type, on becoming active, does not move). When $F_1 = 1$, $F_2 = -40$ and when $F_1 = -40$, $F_2 = 1$, so that in the first case only type-1 cells actively move, in the second only type-2 cells, the final state has the same basic characteristic: the more adhesive cells of type 2 form the inner mass.

Thus when $\alpha = 1$, differences in motility of different cell types have no influence on the nature of sorting out. Also, to obtain sorting out it is sufficient that at least one cell type be capable of active movement.

2. $E^0 = E^\alpha$. In this case the movements of type-i cells are directed toward maximizing E_{ij}^0. If only cells of one type move, the model attempts to reach the configuration corresponding to a maximum of the function E_{ji}^0, $i = 1, 2$; when both cell types move, the model may show cyclic behavior. The value of W_{ij}^k that defines the form of the maxima of E_{12}^0 and E_{21}^0 is chosen, with the help of condition (C) in section 2, in such a way that E_{12}^0 would be at a maximum for a sorting out with type-1 cells in the interior, and E_{12}^0 for a sorting out with type-2 cells in the interior.

On the basis of the discussion in section 2 it is possible to surmise that when $\alpha = 0$, differences in the values of F_1 and F_2 determine the nature of sorting out. To test this a series of experiments was performed using fixed values of W_{ij}^k and $F_2 = 1$, while the value of F_1 was varied from -50 to $+1$. The percentage of aggregated cells was shown to depend weakly on F_1, while at the same time the percentage of type-1 cells in the boundary of the aggregate changed from 100% for $F_1 = -50$ to 50% for $F_1 = 1$.

The final results for these cases are given in figures 4.2a and 4.2b. Sorting out occurred only in the case shown in figure 4.2a. The maximal value of F_1 at which sorting out was still observed is -45, which corresponds to a complete stoppage of type-1 cells.

In a second series the effect of differences in the adhesion energy of different types of cells was tested and was found to be nil. In all cases where sorting out occurred, the inner component was formed by the stopped cells, independent of their adhesiveness W_{21}^k and W_{12}^k.

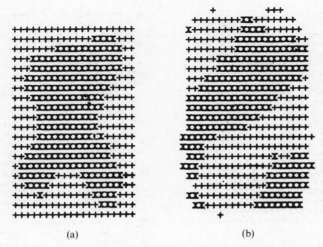

Fig. 4.2. Variant A, $R = 2$, $\rho_1 = \rho_2 = \frac{1}{2}$, dimensions of the aggregate 20×20, state at time $t = 15\tau$. (a) $F_1 = -1000$, "standard" values for variant A; (b) $F_1 = +1$.

Thus, if it is the total energy of adhesion that is maximized, the nature of sorting out is determined by the adhesion energy of the cells and does not depend on the values of F_1 and F_2, that is, on the power of their motility apparatus if only they permit sorting out. If the energy of adhesion is maximized separately for each type of cell, sorting out depends on the differences in mobility of cells and does not depend on their adhesiveness. At present it is not clear which case is closer to biological reality. One can only say that the first variant leads to sorting out in the great majority of initial configurations, which can be seen from the following experiment.

The initial configuration is composed of three types of cells (0, 1, and 2) randomly dispersed on a square with probabilities $\rho_1 = \rho_2 = \rho_0 = \frac{1}{3}$. Taking $\alpha = 1$ and the corresponding W_{ij}^k and F_{ij}, the model reaches a state where cells of types 1 and 2 are compacted together (fig. 4.3a). The outer dimension of the aggregate decreases relative to the initial one because some of the zeros (30%) pull away from the initial loose cluster. The remaining zeros form a cavity in the interior of the aggregate. Both the outer and inner boundaries are composed of the less adhesive, type-1 cells. This sort of structure is similar to the pattern of reconstruction of embryonic lung tissue, if one regards 2 as mesenchyme and 1 as epithelial cells.

When $\alpha = 0$, only cells of type 2 form compact masses, which are surrounded by a layer of translocated zeros and ones (fig. 4.3b). This is natural because the difference between F_1 and F_2, which is essential for sorting out, makes it impossible for type-1 cells to translocate actively. These findings suggest a great generality for models with $\alpha = 1$.

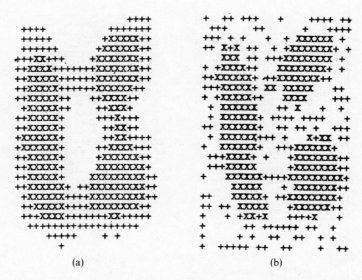

Fig. 4.3. $R = 3$, $\rho_1 = \rho_2 = \frac{1}{3}$; dimensions of aggregate 24×24; state at time $t = 15\tau$. (a) Variant A and (b) variant B.

5. SORTING OUT WHEN $\alpha = 1$ (STEINBERG'S CASE)

From the experiments of section 3 it follows that to obtain the correct pattern of sorting out in small aggregates, it suffices to conform to condition (A). The experiments in this section show that for the majority of the values of the parameters W_{ij}^k there exist regions of sorting out that are not regions of (A). All these experiments concern the case where $\alpha = 1$, $E_{12}^a = E_{21}^a = E$, that is, the movement of cells is accompanied by an increase in the overall adhesion energy of the aggregate, in conformity with the hypothesis of Steinberg. Condition (A) for $\alpha = 1$ is derived below.

Take

$$a_k = W_{11}^k + W_{22}^k - 2W_{12}^k$$

$$a_1 > 0, \qquad a_k \geqslant 0 \tag{1}$$

$$a_k \geqslant a_{k+1} \tag{2}$$

$$0 \leqslant W_{11}^k < W_{12}^k, \qquad W_{11}^k \geqslant W_{11}^{k+1}. \tag{3}$$

Condition (3) guarantees that the internal component will consist of type-2 cells. In order to have the internal component consist of type-1 cells, condition (3) has to be changed to

$$0 \leqslant W_{22}^k \leqslant W_{12}^k, \qquad W_{22}^k \geqslant W_{22}^{k+1}. \tag{3'}$$

Three types of violations of conditions (1)–(3) were studied: a violation of the monotonicity of a_k, a replacement of the condition $a_1 > 0$ by $a_1 = 0$, and (with some supplementary constraints) a violation of the conditions $W^k_{11} \leqslant W^k_{12}$ and $a_k \geqslant 0$.

1. In the violation of monotonicity conditions (1) and (3) hold, but not (2). Figures 3.3b and 5.1 show the final states of the model, which correspond to $a_1 : a_2 : a_3 = 1 : \frac{1}{2} : \frac{1}{3}$ and $a_1 : a_2 : a_3 = 1 : 3 : 5$. Both states are patterns of sorting out and are almost identical. The same observation can be made in the other cases studied: $(1:1:1)$, $(1:1:2)$, $(1:2:1)$, $(1:1:3)$, $(1:3:1)$, $(1:2:3)$, $(1:3:5)$, $(1:5:1)$, and $(1:1:5)$. Thus if constraints (1) and (3) are satisfied, the condition of monotonicity can be relaxed or at least greatly weakened.

Fig. 5.1. Parameters α, W^1_{ij}, and F_i as in variant B, but $W^2_{ij} = 3W^1_{ij}$, $W^3_{ij} = 5W^1_{ij}$ ($a_1 : a_2 : a_3 = 1 : 3 : 5$). Diameter of the aggregate 27; state at time $t = 15\tau$. Compare with figure 3.2b.

2. This case occurs when $a_1 = 0$ was modeled with $W^1_{ij} = 0$, $i, j = 1, 2$; $W^k_{ij} \geqslant 0$ when $k = 1, 2$ (no contacts of length 1). It appears that under these conditions the final state of the model is one of two types, examples of which are shown in figures 3.2b and 5.2. The state depicted in figure 3.2b is observed when $a_2 > 0$, $a_3 > 0$ and is a pattern of sorting out. Figure 5.2 corresponds to $a_2 > 0$ and $a_3 = 0$; as is obvious, the type-2 component has spaces that are filled with type-1 cells. It is not difficult to see that when $a_1 = a_3 = 0$, $a_2 > 0$, the pattern of alternating layers of 1 and 2 is a point of local maximum*

---

* A point of local maximum is a configuration in which any interchange of two neighboring cells of different types leads to a decrease in E.

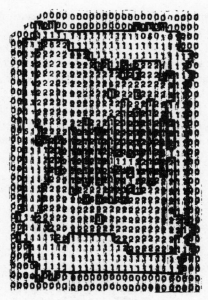

Fig. 5.2. Parameters α, W_{ij}^3, and F_i as in variant B, but $W_{ij}^1 = W_{ij}^2 = 0$. Dimensions of the aggregate 25×25; state at time $t = 15\tau$.

of the function E, which, it would seem, explains the effect. The same result was observed when $a_1 = a_2 = 0$, $a_3 > 0$, but the minimal distance between the holes and their width is equal to two. This relationship is in conformity with the fact that when $a_1 = a_2 = 0$, $a_3 > 0$, the alternating layers 1 and 2 are a point of local maximum of E. Thus when $a_2 > 0$ and $a_3 > 0$, the condition $a_1 > 0$ can be relaxed.

3. Violation of the conditions $a_k \geqslant 0$ and $W_{11}^k \leqslant W_{12}^k$ was modeled by putting certain restrictions on the magnitude of W_{ij}^k. They are:

(a) $W_{ij}^k = 0$ when $k > R(i, j)$, where $R(i, j) = i + j - 1$.

This condition signifies that cells of type 1 have only short bonds, whereas those of type 2 include short ones and also bonds of length 2 and length 3.

(b) We set

$$u_{ij} = \sum_{R=1}^{R(i,j)} k W_{ij}^k$$

and require that

$$u_{11} < u_{12} \tag{4}$$

$$u_{11} + u_{22} - 2u_{12} > 0. \tag{5}$$

Conditions (4) and (5) are obtained from the following qualitative considerations. If in the expression

$$E = \sum_{k=1}^{R(i,j)} \sum_{ij} N_{ij}^k W_{ij}^k,$$

kN'_{ij} is substituted for N_{ij}^k, then E has the form

$$E(u) = u_{11}N'_{11} + u_{12}N_{12} + u_{22}N'_{22}.$$

This expression is maximal at the state of sorting out if conditions (3) and (4) are fulfilled (see section 2). It is easy to show that for sorting out to occur and for configurations with random distributions of cells, $N_{ij}^k \approx kN'_{ij}$.

Below are described two series of experiments, A and B. In series A the value of W_{ij}^k does not depend on k when $k \leqslant R(i,j)$; in series B these values may change with changes in k.

Series A includes three pairs of variants with the following parameter values:

i,j \ k	1	2	3	i,j	1	2	3	i,j	1	2	3
1, 1	4	0	0	1, 1	3	0	0	1, 1	10	0	0
1, 2	3	0	0	1, 2	4	0	0	1, 2	4	0	0
2, 2	4	0	0	2, 2	3	0	0	2, 2	3	0	0
1, 1	4	0	0	1, 1	3	0	0	1, 1	10	0	0
1, 2	3	3	0	1, 2	4	0	0	1, 2	4	4	0
2, 2	4	4	0	2, 2	3	3	0	2, 2	3	3	3

The values of W_{ij}^k in the lower three lines of the table above produce the following results:

1. A redistribution of the initial random configuration into a small number of aggregates consisting of cell types 1 and 2 without the appearance of internal and external components (fig. 4.2b)
2. A structure composed of alternating layers of 1s and 2s (fig. 6.1a)
3. Sorting out (in the given case with an internal component of type 1)

As modeling shows, these are the three basic types of final patterns.

By changing the lengths of bonds (see lower lines of table above) it is possible to obtain changes in the final pattern. It remains unclear whether by changing the lengths of some bonds it is possible to force the cells to produce a pattern of sorting out with an internal component of type 2 instead of a 1–3 pattern. The patterns obtained (corresponding to the lower part of the table) are all of one type (fig. 5.3) and represent a region that is filled by alternating layers of 1s and 2s surrounded by a layer of 1s.

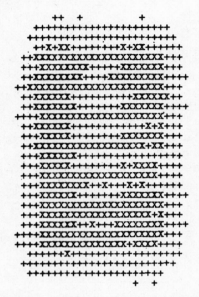

Fig. 5.3. Pattern of sorting out when condition B is violated.

In series **B** an attempt was made to study the effect of relaxing the requirements of the inequalities $a_k \geqslant 0$ and $W_{11}^k \leqslant W_{22}^k$ for each k separately. The possible cases are enumerated in Table 5.1, where plus signifies that the inequality holds, minus that it does not. The case $\{W_{12}^2 < W_{11}^2, W_{12}^3 < W_{11}^3\}$ does not occur in conformity with condition 3(a) (see p. 70).

Sorting out occurred in two cases out of six [disregarding the first line, where conditions (1)–(3) are fulfilled]. In the remaining cases the final state was the same as in figure 5.3. These experiments show that inequalities (1)–(3), fulfillment of which ensures sorting out, can be weakened. For example:

1. The requirement for a monotonic decrease of a_k can be relaxed.
2. The requirement $a_1 > 0$ can be replaced by $a_2 > 0$, $a_3 > 0$.
3. The requirements $a_k \geqslant 0$ and $W_{11}^k \leqslant W_{12}^k$ can be relaxed in certain cases.

6. Modeling of Invasion and of Aggregation

Application 1

The phenomenon of invasion is observed during metastasis of tumors, in which the tumor cells migrate into the surrounding tissues. A structure is formed consisting of layers in contact with one another or of bands of

Table 5.1.

W_{ij}^k	$W_{11}^k \leqslant W_{12}^k$	$\alpha_1 \geqslant 0$	$\alpha_2 \geqslant 0$	Result
1 2 1 4 3 1	+	+	+	Sorting
2 1 1 1 3 1	−	+	+	Sorting
1 2 1 0 3 1	+	−	+	Fig. 5.3
3 2 1 0 3 2	−	−	+	Sorting
1 2 2 4 3 2	+	+	−	Fig. 5.3
1 2 1 1 1 3	+	−	−	Fig. 5.3
2 1 1 3 1 2	−	+	−	Fig. 5.3

tumor cells. The question arises whether it is possible to produce a similar phenomenon within the framework of our model. (The problem of modeling cell invasion was suggested by Vasiliev.)

It should be noted that invasion by a tumor is always accompanied by a multiplication of its cells; also, the cells of a tumor are as a rule anisotropic. This situation cannot be represented in our model. We are also unable to model the migration of tumor cells through and along blood vessels, which is an important aspect of metastasis. Because of this constraint we model only a short period of the dispersion of a tumor of isotropic cells into the surrounding homogeneous tissue.

The simplest assumption is that the intrusion of a group of tumor cells is caused by an increase in their adhesiveness with respect to normal cells that surround them. It is possible to show (Goel et al.[7]) that for values of the adhesion energy that conform to the relations

$$W_{12} > W_{22}, \quad W_{12} > W_{11},$$

a configuration of cells having the maximum energy of adhesion has a complex structure in which each cell forms the maximum number of bonds with cells of another type.

The results of modeling when $W^k_{11} = W^k_{22} = 1$, $W^k_{12} = 3$, and $k = 1, 2, 3$ are displayed in figure 6.1a. The initial configuration of the aggregate was two aggregations of type-1 and type-2 cells on a grid of size 20×10, in contact with each other along the longer size. Figure 6.1b shows an inter-

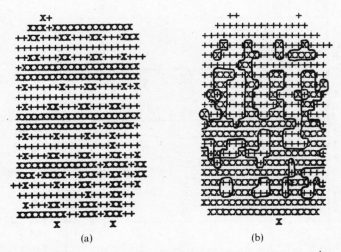

(a)   (b)

Fig. 6.1. (a) and (b). Invasion for the case of long bonds between cells. $\alpha = 1$, $W^k_{11} = W^k_{22} = 1$, $W^k_{12} = 3$, $k = 1, 2, 3$; $F_i = +1$. Size of the aggregate 20×20.

Fig. 6.1a. State at time $t = 10\tau$. Fig. 6. 1b. The development of invasion. State at time $t = 3\tau$.

mediate stage of intermixing at time 3τ. The final pattern consists of alternate layers of different cells. Subsequent experiments showed that the width of the layers decreases with a decrease in the length of bonds between the cells (fig. 6.2a).

The progress of intermixing with time results in the formation of ever-longer ribbons of cells extending into the mass of cells of the opposite type and in separation from the main mass and from the ribbons of separate groups of cells. The time for stabilizing the pattern is not longer than 5τ.

When there are differences in the bond lengths of different types of cells there appears in the aggregate a zone of mixing and a zone of aggregation, which form the boundary of the main cell mass (fig. 6.2b).

The regular layered structure obtained in the experiments is not the result of a regular initial configuration: with a random initial configuration the

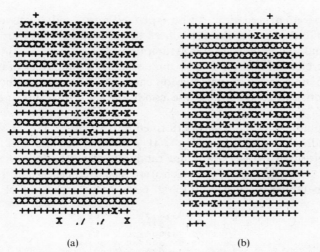

(a)   (b)

Fig. 6.2a. Invasion with short bonds between cells. Parameters the same as in figure 6.1, but $k = 1$.

Fig. 6.2b. Pattern of invasion with different length of bonds between cells. $\alpha = 1$, $W_{11}^1 = 1$, $W_{12}^1 = W_{12}^2 = 3$, $W_{22}^1 = W_{22}^2 = W_{22}^3 = 1$, the other $W_{ij}^k = 0$, $F_1 = F_2 = +1$. State at time $t = 15\tau$.

same phenomenon was observed in even more marked form. Apparently this occurrence is related to the strict homogeneity of bonds in the model.

According to biologists who have seen the patterns involved, only the case with long bonds is similar to a picture of invasion (fig. 6.2a). They find that intermixing in the model resembles the real case fairly closely.

Although these findings were obtained with a model that greatly simplifies real invasion, one gets the impression that this phenomenon can be explained in principle as the result of the same rules of behavior that occur in sorting out.

Application 2

The following experiments pertain to the problem of the reconstruction of lower animals (sponges and *Hydra*), to which many investigations have been devoted. A series of findings indicates that aggregations of the cells of a sponge, which is the precursor stage to reconstruction, is species specific.[10,11]

In a typical experiment the cells of two species of sponges obtained by tissue dissociation are allowed to settle to the bottom of a vessel. The cells move actively and come into contact. Some of these contacts result in adhesion, so that one observes aggregates of cells on the bottom of the vessel, each of which consists of cells of the same species. Reconstruction then proceeds and leads to the complete organization of a sponge.

The separate aggregation can be explained most easily by assuming that cells of different species do not adhere to each other. There are data that support and that contradict this assumption.

It is therefore of interest to determine whether simple rules of movement, leading to an increase in the adhesion energy of cells, can produce separate aggregation and to determine the conditions necessary for specificity of adhesion.

A computer experiment was performed for two cases: $W_{11} > W_{12} < W_{22}$ (specific adhesion) and $W_{11} < W_{12} < W_{22}$ (nonspecific adhesion). The model for the bottom of a vessel containing randomly dispersed cells of two species was a 50×50 grid region, on which were dispersed type-1 and type-2 cells with probabilities $\rho_1 = \rho_2 = 0.1$ (fig. 6.3a shows a portion of the initial

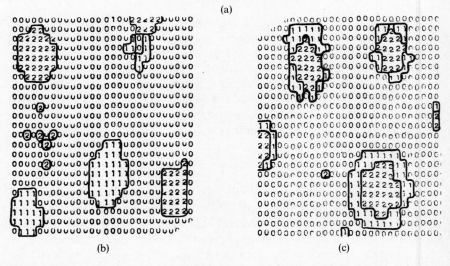

(a)

(b)

(c)

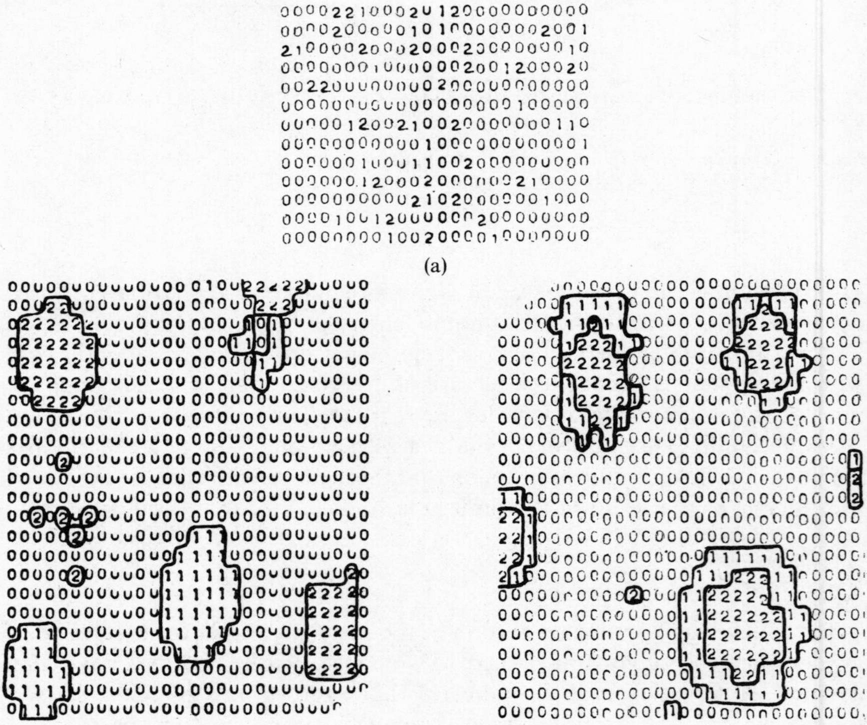

Fig. 6.3a. A portion of the initial state of the aggregate of size 50×50 with $\rho_1 = \rho_2 = 0.1$.

Fig. 6.3b. Aggregation with species specific adhesion. $W_{11}^k = W_{22}^k = 1$, $W_{12}^k = 0$, $k = 1, 2, 3$, $F_i = +1$. State at time $t = 15\tau$.

Fig. 6.3c. Aggregation with nonspecific adhesion. $W_{11}^k = 1$, $W_{12}^k = 3$, $W_{22}^k = 6$, $k = 1, 2, 3$, $F_i = +1$. State at time $t = 15\tau$.

configuration). The results are given in figures 6.3b and 6.3c. In each case aggregation of cells occurred, but when $W_{11} > W_{12} < W_{22}$, aggregation occurred separately for the two species. In the nonspecific case, mixed aggregates appear in which sorting out takes place. Repeated experiments with $\rho_1 = \rho_2 = 0.05$ and with varying numbers of bonds per cell produced findings that were qualitatively the same.

Apparently, even if there is a mechanism for separate aggregation that operates under conditions of nonspecific adhesion, it must operate on principles not based on the increase in the overall energy of bonds.

REFERENCES

1. Moscona, A. A. (1957) The development in vitro of chimaeric aggregates of dissociated embryonic chick and mouse cells. *Proc. Nat. Acad. Sci. U.S.* 43: 184–94.
2. Steinberg, M. S. (1964) The problem of adhesive selectivity in cellular interactions. In *Cellular Membranes in Development*, ed. M. Locke, pp. 321–36. New York: Academic Press.
3. Curtis, A. S. G. (1960) Cell contacts: some physical considerations. *Amer. Naturalist* 94: 37–56.
4. Trinkaus, J. P. (1961) Affinity relationships in heterotypic cell aggregates. La Culture Organotypique. *Colloq. Int. Centre Nat. Rech. Sci. Paris* 101: 209–26.
5. Trinkaus, J. P. and Lentz, J. P. (1964) Direct observation of type-specific segregation in mixed cell aggregates. *Develop. Biol.* 9: 115–36.
6. Trinkaus, J. P. (1967) Morphogenetic cell movements. In *Major Problems in Developmental Biology*, pp. 125–76. New York: Academic Press.
7. Goel, N., Campbell, R. D., Gordon, R., Rosen, R., Martinez, H., and Yčas, M. (1970) Self-sorting of isotropic cells. *J. Theor. Biol.* 28: 423–68.
8. Leith, A. G. and Goel, N. (1971) Simulation of movement of cells during self-sorting. *J. Theor. Biol.* 33: 171–88.
9. Vasiliev, A. V. and Pyatetskii-Shapiro, I. I. (1971) Modeling of the process of cell sorting on a computer. *Ontogenez* 2: 356–61.
10. Wilson, H. V. (1907) On some phenomena of coalescence and regeneration in sponges. *J. Exp. Zool.* 5: 245–58.
11. Curtis, A. S. G. (1970) Re-examination of a supposed case of specific cell adhesion. *Nature* 226: 260–61.

5. On a Problem of Sorting Out

A. V. Vasiliev*

The following problem is considered. In each cell of a whole-number grid there is an element that is of one of two types. At each step a pair of differing neighboring elements is chosen at random, which then exchange places with a probability depending on the type of neighbors. The necessary and sufficient conditions are found for the existence of a final measure of the Gibbsian type for an arbitrary grid with an arbitrary number of elements.

The sorting of randomly mixed elements of several different types is observed in a number of physical and biological processes; see, for example, Ebert.[1] This paper concerns the problem of modeling these processes in homogeneous random networks, a problem posed by Pyatetskii-Shapiro. The definition of a homogeneous random network was formulated by Stavskaya and Pyatetskii-Shapiro.[2] Knowledge of the work of Shmukler[3] proved very useful.

We examine a whole-number grid of dimensions $n = 1, 2,$ or 3[†] with whole-number coordinates $(x_1, \ldots, x_n)$ and a region of it that consists of all nodes of the grid with coordinates $0 \leqslant x_1 \leqslant m_i, i = 1, \ldots, n$. We shall assume that opposite sides of this region (i.e., the set of nodes with coordinates $x_i = 0$, and $x_i = m_i, i = 1, \ldots, n$) are pairwise identified. We shall call such a region with pairwise-corresponding oppositely situated sides a *whole-number torus* and shall denote it by $T^n(m_1, \ldots, m_n)$ or T^n.

At each node of the torus $T^n(m_i, \ldots, m_n)$ let there be exactly one element of type 1 or 2. At the time t $(t = 1, 2, 3, \ldots)$ a pair of neighboring nodes is chosen randomly and with equal probability. (In the torus T^n each node has $2n$ nodes neighboring a given coordinate direction.) With a certain probability there is an exchange of elements in the given chosen nodes. It is assumed that if the pair selected consists of identical elements, this probability is 0; in the opposite case it is equal to π_{ij}, $0 \leqslant \pi_{ij} \leqslant 1$, where i is the number of type-1 neighbors and j is the number of type-2 neighbors.

Such a procedure produces a certain Markov chain. We denote the total number of elements of type 1 in T^n by N_1 and the Markov chain by $M(T^n, N_1)$.

* First published in *Problemi Peredachi Informatsii* (*Problems of Information Transfer*) (1971), no. 3: 109–11.
† We limit ourselves to the physical case of $n = 1, 2,$ or 3, but theorems 1 and 2 can be shown to hold for n.

This chain's set of states is the set of the distribution of the type-1 and type-2 elements on the torus T^n; let us call this set $\Omega(T^n, N_1)$.

In our calculations we shall find an invariant Gibbsian measure of the chain $M(T^n, N_1)$ for the particular case $\pi_{ij} = \pi_{i+j} = \pi_k$. This measure proved to have the same form as the measure postulated in the model of Ising.

Let ω be an arbitrary element of Ω. We shall denote the number of pairs of neighboring elements of different types by $\Gamma(\omega)$; therefore $\Gamma(\omega)$ is the length of the boundary between masses composed of type-1 and type-2 elements in state ω. Also let $\pi_{ij} = \pi_{i+j} = \pi_k$.

Theorem 1. (a) If $\pi_k = ca^k$, $1 \geqslant c \geqslant 0$, $ca^{4n-2} \leqslant 1$, then the $\mu(\omega) = ca^{\Gamma(\omega)}$ is an invariant measure of chain $M(T^n, N_1)$ for any N_1. (b) There exists such a natural number $a(n)$, $a(n) > 1$, that if the dimensions of the torus are $m_1, \ldots, m_n$, a multiple of $a(n)$, then from the invariance of the measure $c_1 a^{\Gamma(\omega)}$ for any N_1, it follows that π_k satisfies condition (a).

Proof (a). Sufficiency. Let $\pi_k = ca^k$ and intermediate probability values of the chain be denoted by $p(\omega', \omega)$. To verify the equation

$$\mu(\omega) = \sum_{\omega' \in \Omega} p(\omega', \omega)\mu(\omega'), \qquad (1)$$

where $\mu(\omega) = c_1 a^{\Gamma(\omega)}$, $\omega \in \Omega$, we note that the sum is taken only for ω', in the set $R(\omega)$ which consists of ω and those elements of Ω differing from ω by one exchange.

We shall say that a pair of different neighboring elements has the index k if $i + j = k$; here, as above, i and j are the corresponding numbers of type-1 and type-2 neighbors of elements 1 and 2, respectively. As can be seen, an exchange between a pair of elements of index k leads to a pair with index $4n - 2 - k$; in addition, if before exchange the configuration had a boundary of length Γ, after exchange within the pair of index k the length of the boundary becomes equal to $\Gamma + 4n - 2(1 + k)$.

If ω has $r_k(\omega)$ pairs with index k, then $r_k(\omega)$ states, which differ from ω by one exchange within a pair of index $4n - 2 - k$, enter into the set $R(\omega)$. Let ω'_k be any of these states. It is clear that ω'_k does not transform into ω by any exchange of another pair of any index and that the relation

$$p(\omega'_k, \omega) = \frac{1}{N} \pi_{4n-2-k}$$

$$\mu(\omega'_k) = c_1 a \exp\left[\Gamma(\omega) - 4n + 2(i + k)\right]$$

is correct (here N is the total number of pairs of neighboring nodes in T^n).

Further, the probability $p(\omega, \omega)$ is composed of the probabilities of choosing the same elements and of choosing a pair of different elements and not completing the exchange. From this operation,

$$p(\omega, \omega) = 1 - \frac{1}{N} \sum_{k=1}^{4n-2} r_k(\omega) + \frac{1}{N} \sum_{k=1}^{4n-2} r_k(\omega)(1 - \pi_k) = 1 - \frac{1}{N} \sum_k r_k(\omega)\pi_k.$$

Equation (1) takes on the form

$$a^{\Gamma(\omega)} = \left[1 - \frac{1}{N} \sum_k r_k(\omega)\pi_k \right] a^{\Gamma(\omega)} + \frac{1}{N} \sum_k r_k(\omega)\pi_{4n-2-k} a$$

$$\times \exp\left[\Gamma(\omega) + 2(1 + k) - 4n\right]$$

or

$$\sum_k r_k(\omega)\left[\pi_k - \pi_{4n-2-k} a^{2(1+k-2n)}\right] = 0.$$

The last relationship is evident from the condition (2) imposed on π_k.

(b). Necessity. Let Equation (1) hold for arbitrary N_1 and T^n with sizes $m_1, \ldots, m_n$, a multiple of some $a(n)$. The quantity $a(n)$ will be selected during the course of the proof. We shall assume that $n = 2$ (for $n = 1$ and $n = 3$ the proof is analogous).

We shall demonstrate that there is a choice of states $\omega_0, \ldots, \omega_6$ on a torus T_n for which $r_k(\omega_k) > 0$, $r_k(\omega_i) \geqslant 0$ when $k \leqslant i$; $r_k(\omega_i) = 0$ when $k > i$, $i = 0, \ldots, 6$. The numbers of type-1 and type-2 elements in these states may be different. In that case the system of equations

$$\sum_k r_k(\omega_i) y_k = 0, \qquad i = 0, \ldots, 6,$$

which is equivalent to equation (1) by Gibbsian measure, has a unique solution $y_k = 0$, $k = 0, \ldots, 6$, from which condition (a) of theorem 1 follows.

Shown below is one of the possible choices $\{\omega_i\}$. Each ω_i is doubly periodic, that is, there are two shifts of the state on the grid, along $l_1(\omega_i)$ in the coordinate direction x_1 and along $l_2(\omega_i)$ in the coordinate direction x_2, which transform ω_i into itself. For each ω_i, the distribution of elements is indexed in the rectangle

$$1 \leqslant x_1 \leqslant l_1(\omega_i), \qquad 1 \leqslant x_2 \leqslant l_2(\omega_i).$$

$i = 0$:	12	$i = 1$:	2121	$i = 2$:	112	$i = 3$:	112
	21		1112		121		121
			2121		211		221
			1212				

$$i = 4: \quad 12 \qquad i = 5: \quad 112 \qquad i = 6: \quad 1122$$

It can be seen that the quantity $a(2)$ can be chosen to equal 12. For another choice of the state $\{\omega_i\}$, the value of $a(2)$, generally speaking, will change.

Theorem 1 can be generalized as follows. We shall suppose that some nodes of the grid are not filled in with elements. Also, we shall introduce into π_k the probabilities of exchanges 1—0 and 2—0 and shall denote them by ρ_{1l} and ρ_{2l}, respectively, where l is the increase in the number of type-1 neighbors (type-2 neighbors) of 1 (2) elements that results from its moving into an empty node. The increase may also be negative. We shall also denote the total number of type-2 elements by N_2, and the number of pairs (1—0) and (2—0) in the state ω by $\Gamma_1(\omega)$ and $\Gamma_2(\omega)$, respectively.

Theorem 2. The measure $v(\omega) = ca_1{}^{\Gamma_1(\omega)} a^{\Gamma_2(\omega)} (a_1 a_2)^{\Gamma(\omega)}$ is an invariant measure of the chain for arbitrary N_1, N_2, and T^n if

$$\pi_k = [c(a_1 a_2)]^k, \qquad \rho_{1l} = c'a_1^l, \qquad \rho_{2l} = c''a_2^l.$$

This theorem is proved analogously to theorem 1.

REFERENCES

1. Ebert, J. D. (1965) *Interacting Systems in Development*. New York: Holt, Rinehart and Winston.
2. Stavskaya, O. N. and Pyatetskii-Shapiro, I. I. (1968) On homogeneous networks of spontaneously active elements. *Probl. Kibernetiki* (*Probl. Cybernetics*) 20 M: Fizmatgiz, 91–106.
3. Shmukler, Yu. I. (1968) A thermodynamic model of adaptation. *Dokl. Akad. Nauk SSSR* 182: 6, 1290–93.

6. Reconstruction of Tissues by Dissociated Cells

Malcolm S. Steinberg*

How is the structure of a multicellular animal generated? In the broadest terms, we can distinguish three kinds of developmental processes: growth, differentiation, and morphogenesis. The developing organism multiplies its cells and increases its mass. The emergent parts become different—different from what they were before and different from one another. And the differentiating parts bend inward or outward, expand, contract, disperse, condense, fuse, separate, elongate, even perish, and otherwise rearrange themselves in the process of constructing the animal. But what are the mechanisms that elicit and orient these tissue movements of morphogenesis?

BACKGROUND OF THE PROBLEM

Early workers envisioned the tissue movements as resulting from pressures or other inhomogeneities in the immediate environment, but a considerable body of evidence has meanwhile been accumulated to show that the movements are due to intrinsic properties of the individual tissues themselves. Beyond this statement, however, we find ourselves in an area of uncertainty because the character of these intrinsic properties has not been securely and rigorously established. A crack in the shell surrounding this problem appeared very early. Wilson discovered in 1907 that the cells and cell clusters obtained by squeezing a sponge through the meshes of fine silk bolting cloth could reunite and that aggregates obtained in this way could reconstitute themselves into functional sponges.[1] The manner in which this reconstitution was effected remained problematical. Wilson continued to maintain[2] that a considerable amount of dedifferentiation and redifferentiation occurred and that cells altered their cytological characteristics to conform with their newly established environments, whereas other workers[3] believed they had demonstrated the reconstitution to consist, in large measure, of a sorting out of the various types of cells, each coming again to occupy its accustomed haunts in the body of the sponge. The difficulty lay in the absence of permanent

* First published in *Science N.Y.* (1963) 141: 401–8. Copyright 1963 by the American Association for the Advancement of Science.

and recognizable characteristics by which one could accurately distinguish and follow the various types of cells during the process of reorganization.

In the meantime, Harrison[4] had laid the foundation for modern neuro-embryology, a foundation that included the concepts of the selection of paths by outgrowing nerves and of the specificity of nerve-end organ connection, and that was ably extended and built upon by the researches of Weiss,[5] Hamburger, and others.

A second discovery of major importance appeared against this background in 1939. Holtfreter, working with carefully defined tissue fragments from young amphibian embryos, found that these fragments showed marked preferences in their adhesive properties. These preferences were correlated with their normal morphogenetic functions. For example, ectoderm and endoderm, isolated from a gastrula, would adhere to each other much as they do at the same stage *in vivo*. In time, however, these two tissues would separate from each other, an event that occurs in the embryo as well. This separation is accomplished in normal development by the penetration of the mesoderm between the ectoderm and endoderm. Mesoderm incorporated along with the isolated ectoderm and endoderm was indeed found to bind the latter two tissues together in a permanent union *in vitro* as it does *in vivo*. Furthermore, when the tissues were present in the right proportions, the ectoderm would take up an external position and the endoderm an internal position, with the mesoderm spread out in between, duplicating in the culture vessel not only the associations but also the anatomical relations that exist in the embryo. An impressive array of similar results with these and other tissues[6] led Holtfreter to frame the concept of "tissue affinities" to describe these associative preferences, which he had shown to be so closely related to normal morphogenetic events.

A third advance was made by Holtfreter in 1944. He found that by subjecting a fragment of an amphibian gastrula to an environmental *p*H of about 10, he could cause the individual cells to separate and fall away from one another, much as Herbst had earlier been able to cause the separation of sea urchin blastomeres in calcium-free seawater.[7] Upon return to a more neutral *p*H, the amphibian cells would reestablish mutual adhesions, attaching themselves to any neighbors with which they came into contact, and building in this manner masses of tissue into which cells of the various germ layers were incorporated at random. The situation resembled that in the sponges but with one important distinction. Differences in the degree of pigmentation of the amphibian cells, together with their extraordinarily large size, allowed the investigator to follow the movements at least of the surface cells. Before his eyes the lightly pigmented mesoderm cells vanished into the depths of the tissue mass, while darkly pigmented ectoderm cells and the almost pigment-free endoderm cells emerged to replace them at the periphery.[8] Sorting out was a reality. And the tissue affinities that Holtfreter had earlier described could with justice be renamed *cell* affinities, for it was now clear that they were inherent in the individual cells.

Other workers made significant contributions. Principal among them was Moscona, who opened the way for investigations with the cells of older avian embryos through his discovery that trypsin was effective in dissociating their tissues.[9] Through the use of this technique, Moscona[10] and Trinkaus and Groves[11] showed that the reconstitution of body parts by aggregates of intermixed cells occurred even though the constituent cells were in all likelihood determined with respect to their fates, and even though they had already reached their appropriate positions within the embryo. The same fact was established by Townes and Holtfreter[12] for older amphibian embryos as well. The remarkable degree to which normal structure could be approximated by a self-organizing cell mixture was demonstrated by Weiss and Taylor,[13] who, by culturing aggregates derived from highly differentiated organs in a site that provided vascularization, obtained organogenesis that strikingly approached the complexity of normal organization.

The foregoing historical account is only a sketch that makes no pretense of complete coverage. It serves, however, to document the fact that formerly elusive problems concerning the mechanisms of morphogenetic movement have been brought more closely within the experimenter's grasp.

SORTING OUT, ADHESION, AND MOTILITY

The most fundamental facts concerning tissue reconstruction are perhaps the following. (1) When the cells of different vertebrate embryonic tissues are dissociated and mixed, they are capable of establishing adhesions with one another and constructing common aggregates. (2) Within such mixed aggregates containing cells from different tissues, the differing kinds of cells regroup, each with the others allied to it, to reconstruct the various tissues of origin. (3) These tissues are reconstructed in definite positions (see also ref. 11); for example, muscle is always built external to cartilage, never the other way around. (4) When the tissues employed are parts of a complex within the embryo, the geometry of the entire normal complex is reflected in the reestablished structures.

In normal development, tissue *X* may spread from some previous position to cover the surface of tissue *Y*. In a mixed aggregate *in vitro*, the same ultimate geometry would be achieved through the sorting out of the jumbled *X* and *Y* cells. The fact that the specific anatomical structure is established by pathways that differ so greatly in the two cases is to be regarded less as a curiosity than as a stroke of great fortune for the student of morphogenesis. It indicates that the features responsible for the ultimate anatomical organization are common to these two disparate systems. In the case at hand, two common features at once come to mind. They are the basic cellular properties of mutual adhesiveness and motility. It is not my purpose here to cover the extensive literature concerning cellular adhesion and cell movement, much of which is discussed in recent publications.[14,15] I wish rather to examine two particular assumptions that, either singly or in combination, are widely held to be necessary in order to account for sorting out and tissue reconstruction. These assumptions are (1) that the segregating cells exhibit actively directed movements and (2) that they display qualitatively selective mutual adhesion.

The segregation of cell species that takes place within mixed aggregates could, in principle, be brought about in either of two ways. Either the differing cells might seek out, by active and directed migration, different parts of the aggregate (or even one another), or they might possess type-specific differences in adhesiveness by virtue of which the old cellular alliances are again progressively built up through the agency of random collisions. Both possibilities, as well as a combination of the two, have been suggested by various authors. In view of the early experimental documentation by Holtfreter[6,16] of differences in adhesiveness among such cells, and because of the apparent cellular selectivity involved in wound healing and in neurogenesis, most of the speculation has centred around possible mechanisms by which adhesion might be rendered selective. There is no body of evidence for mutual attraction (or repulsion) by embryonic cells.

It has been variously proposed that embryonic cells selectively adhere to one another by means of binding sites that possess singularities of conformation,[5,17,18] of chemical composition,[19] or of geometric arrangement[5,14,17,20]; that adhesion among differing cells is nonselective in character but varies in its intensity as a function of cell type and time[21] or as a result of selective influences that favor disjunction[22]; and that in addition to,[12] or possibly in lieu of,[23] showing selectivity in their adhesion to one another, cells may migrate in a directed fashion either inward or outward within multicellular masses, the migration ultimately bringing about their mutual segregation.

In virtually all the hypotheses that have been advanced, attention has been focused upon the adhesive and motor properties of the segregating cells; for without adhesion there can be no coherent multicellular aggregate, and without motility on the part of the component cells there can be no sorting out. Almost all authors who have dealt with this problem (with the exception of Stefanelli et al.[23]) have also assumed that the differences in adhesiveness among the different types of cells are type specific, at least throughout the period during which sorting out occurs. Ample justification for this assumption is to be found in the experimental literature, as I have pointed out. Beyond this point, each additional assumption increases the risk of error.

I wish now to develop the thesis that the behavior that is characteristic of cells in the process of sorting out and tissue reconstruction follows directly from their possession of motility and quantitative differentials in adhesiveness, unrestricted by any requirement for qualitative specificity. It will be helpful in this analysis to review first the behavior of inanimate physical systems that share with living cells precisely these attributes, and to examine the way in which this behavior is influenced by the particular quantitative adhesive relationships that apply. In this way we may see the consequences of the presence of these motor and adhesive properties unobscured by any of the complex and often seemingly goal-directed activities of which cells are capable.

In the physical world, we recognize that the units that comprise a gas are mobile but not coherent—they fly apart to fill as much space as is provided. When the energy that drives them apart is sufficiently reduced, attractive forces begin to dominate and the units form a different type of system—a liquid, in which they retain mobility but gain coherence. Reduction of the thermal energy to a still lower point results in the domination of attractive forces to such an extent that mobility of the units is effectively inhibited, and we have a system in which coherence is retained but mobility is severely restricted—a solid. Thus, in the world of molecules, a liquid system is one that is composed of a population of coherent, mobile units.

Many of the properties of liquid systems depend exclusively upon this fact. It is of no substantive consequence that the units happen to be molecules and that their motility happens to be passive rather than active in nature. These properties are independent of the composition of the units, independent of the causes of their motility, and independent of the nature of the adhesive forces. For example, a liquid drop assumes a spherical shape when subjected to uniform external conditions because the mobile units of which it is composed attract or adhere to one another until the greatest possible number have the maximum possible contact. Adhesion being nothing more than close-range attraction, the same holds true for a population of actively motile, uniformly adhesive cells (fig. 1).

The same principle can be expressed by saying that the *free energy* of the drop reduces to a minimum. Included in this quantity is the *surface free energy*, which provided the impetus in the simple illustration given. The surface free energy is merely the energy available for adhesion but left over in the surface, where adhesions could be formed but have not been. It is readily seen to be directly proportional to both (1) the area of exposed surface and (2) the free energy per unit of surface area, the latter quantity being a direct reflection of the adhesiveness of the units that comprise the surface. The free energy is a potential energy that will tend spontaneously to decrease toward a minimum in *any* population of mobile, coherent units. At this minimum the system is in thermodynamic equilibrium.

Let us now consider the manner in which thermodynamic equilibrium is achieved in a coherent population consisting of two different kinds of mobile

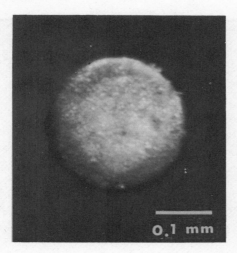

Fig. 1. An initially jagged fragment of liver that has assumed a spherical shape. Isolated from a 5-day chick embryo, it has been maintained in liquid medium at 37°C for 2 days under constant gyration. The same result is obtainable in a stationary culture.

units that adhere to one another with different strengths. The standard measure of the strength of such adhesions is called the work of adhesion. This is a measure of the work done by the system in the formation of an adhesion over a unit of area. (In common usage, the term *work of* ad*hesion* refers to adhesion between two different phases, whereas the equivalent term *work of* co*hesion* refers to adhesion among the units of a single phase.[24]) The units adhere to one another, rearranging themselves as in our first example, until the free energy of the system is reduced to a minimum. This minimum is achieved when the total work done through adhesion in the system is raised to a maximum—in other words, when all the individual units are mutually oriented in such a manner that they adhere to one another with the greatest average tenacity. At this point of thermodynamic equilibrium, the distribution of the two different types of units (phases) within the system is a function of the work of cohesion of each of the two phases and of the work of adhesion between them. There are three types of distribution, each corresponding to one of the following three sets of adhesive relationships, in which the two kinds of units are denoted, respectively, as a and b: (1) a–b adhesions equaling or exceeding in strength the average of a–a adhesions plus b–b adhesions; (2) a–b adhesions weaker than this average but equaling or exceeding in strength the weaker of the other two kinds of adhesions; and (3) a–b adhesions weaker than either the a–a or the b–b adhesions. Let us now explore the three situations to which these relationships correspond.

We will designate the work of cohesion among the units of type a as W_a, the work of cohesion among the units of type b as W_b, and the work of

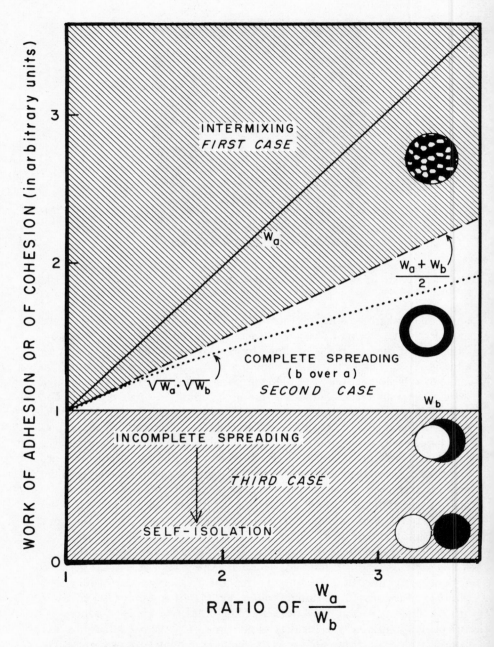

Fig. 2. Types of phase distribution, at equilibrium, in coherent populations consisting of mobile units of two kinds. The work of cohesion of the weakly cohesive *b* units, arbitrarily assigned a value of 1, is given by the line W_b. The work of cohesion of the more strongly cohesive *a* units is denoted by W_a.

adhesion between a and b units as W_{ab}. If it happens that a–b adhesions equal or exceed in strength a value obtained by averaging the strengths of a–a adhesions and b–b adhesions, we can describe the situation by the relation

$$W_{ab} \geqslant \frac{W_a + W_b}{2}. \tag{1}$$

In such a case the greatest average tenacity of adhesion is achieved when a and b units are alternately arranged in the coherent population so as to have the maximum possible interconnection. Therefore, at thermodynamic equilibrium, the two populations are intermixed. This is our case 1 (see fig. 2). If, on the other hand, the strength of the a–b adhesions falls below this average value, the situation is described by the opposite relation:

$$W_{ab} < \frac{W_a + W_b}{2}. \tag{2}$$

In this case the greatest average tenacity of adhesion is achieved when a and b units are totally segregated in the population. However, even the mutual disposition of the segregating a and b phases is thermodynamically controlled, a fact that is shown as follows.

To begin with, let us establish the convention that when the cohesiveness of the units comprising the two phases differs, the more cohesive units will be designated a and the less cohesive units b. Now, if b units adhere to a units with a tenacity that is equal to or greater than the tenacity with which they adhere to one another, we can express this by the relation

$$W_{ab} \geqslant W_b. \tag{3}$$

Relations (2) and (3), taken together, determine the complete set of conditions:

$$\frac{W_a + W_b}{2} > W_{ab} \geqslant W_b. \tag{4}$$

These conditions can only be met when $W_a > W_{ab}$. What are the consequences for the mutual disposition of the segregating phases? Because a–b adhesions are intermediate in strength between a–a and b–b adhesions, the two kinds of units adhere relatively strongly, and the whole coherent population tends to

---

The diagram is used as follows. For any set of adhesive relationships, a vertical line is drawn that passes through the calculated value of W_a/W_b as read on the abscissa. The work of adhesion of a units to b units (W_{ab}), as read on the ordinate, is then entered upon this line. The background shading at this point indicates the distribution of the a and b phases for this system at thermodynamic equilibrium. Example: If $W_a = 3$, with W_b defined as 1, then $W_{ab} = 2.1$ would yield intermixing; $W_{ab} = 1.5$ would yield complete coverage of a by b; and $W_{ab} = 0.5$ would yield incomplete coverage of a by b. The intersection of the vertical line with the dotted line $(W_a)^{1/2}(W_b)^{1/2}$ marks the value of W_{ab} that would be generated in the model system devoid of adhesive specificity, as described in the text (modified from Steinberg[29]).

assume a spherical form in which the exposed surface area is minimal. However, the surface free energy of the system is the product of the exposed surface area and the free energy/unit of such area, and the free energy/unit area, as we saw earlier, is a measure of the cohesiveness of the units in the surface. Consequently, minimization of the surface free energy is achieved only when the surface is of minimal area and contains exclusively the less cohesive of the two kinds of units that comprise the population. Therefore, in the segregation of the *a* and *b* phases, phase *b* will come to occupy completely the surface of the system, which will as a whole tend to assume spherical form. Furthermore, the greatest possible segregation of the two phases will occur, a condition that requires that the interfacial area between the two be minimized. Phase *a* being totally subsurface, this condition is met when phase *a* itself assumes the form of a sphere totally enclosed by phase *b*. Thermodynamic equilibrium is thus established when the less cohesive units are arranged as a coherent sphere totally enclosing a second sphere composed of all the more cohesive units. This is our case 2 (see fig. 2).

Only one other set of possible adhesive relationships remains at this point to be explored. What will be the disposition of the phases at thermodynamic equilibrium if *a–b* adhesions are the weakest adhesions, instead of being as strong as or stronger than the average of *a–a* and *b–b* adhesions (as in our case 1), or weaker than this average yet as strong as or stronger than *b–b* adhesions (as in our case 2)? This circumstance is described by the relationship

$$W_a \geq W_b > W_{ab}. \tag{5}$$

Let us begin by considering the most extreme possible examples. At the one extreme, *a* and *b* units do not adhere to one another at all. Clearly, two separate, isolated populations will form. Each will consist at equilibrium of a sphere containing one of the two types of units. At the other extreme, *a* and *b* units adhere to one another with a strength ever so slightly less than that achieved by the adhesion of a pair of *b* units to each other. Were the *a–b* adhesions any stronger, they would be equal in strength to the *b–b* adhesions and we would have at equilibrium the limiting example of case 2: a sphere within a sphere. Instead, the distribution of the phases at equilibrium is shifted slightly from this configuration in the direction of isolation of the phases: phase *b* recedes at one spot, exposing a minute area of phase *a* at the periphery. The lower the strength of the *a–b* adhesions, the greater the recession, at equilibrium, of the margins of phase *b* around the spherical perimeter of phase *a*. This is our case 3 (see fig. 2). This type of circumstance is classically described by the relationship known as Young's equation.[24] Expressed in terms of work of adhesion, this equation becomes

$$2\frac{W_{ab}}{W_b} = 1 + \cos \theta, \tag{6}$$

where θ represents the internal angle of contact, at equilibrium, of the margin of phase b with the surface of phase a. It has recently been proved[25] that Young's equation is a "direct consequence of: (1) the existence of interfacial free energies and (2) the total free energy of a system at equilibrium being a minimum." The various possible relationships among W_a, W_b, and W_{ab}, and the topographic relationships that they engender, are shown diagrammatically in figure 2. They may be derived from equations presented in most standard texts on surface chemistry (see, for example, ref. 24).

PHASE REDISTRIBUTION IN COHERING POPULATIONS OF EMBRYONIC CELLS

The regroupings discussed in the foregoing section tend spontaneously to occur, for thermodynamic reasons, within a population of mobile, mutually adhesive units of two kinds when the latter are brought into contact. Vertebrate embryonic cells of different kinds are both mobile and mutually adhesive, and they tend, when mixed, to regroup themselves in a manner that often resembles the regroupings described in cases 2 and 3. It is of considerable interest, therefore, to inquire in what measure such sorting out, with its anatomically precise consequences, may be explained by the thermodynamic considerations that have been outlined. Precise measurement of the work of cohesion between living cells does not as yet appear to be feasible. However, it has proved possible to examine in some detail the behavior of mixed populations of embryonic cells and to compare the observed behavior with that to be expected on thermodynamic grounds from a system conforming with one or another set of interunit adhesive relationships.

Chick-embryo heart and neural retinal cells, when mixed in appropriate proportions and allowed to coaggregate in a culture vessel, sort out to form islands of heart tissue totally encased by retinal tissue (fig. 3).[26,27] The system in this respect resembles that in case 2, brought about by conformance with the adhesive relationships given in relation (4). The heart cells correspond with a, and the retinal cells with b. If these adhesive relationships actually operate to bring about the conformation depicted in figure 3, several predictions from thermodynamic theory ought to be fulfilled by the behavior of appropriate mixtures of such cells.

Prediction 1. The replacement of heart cells by retinal cells in the surfaces of heart-retina aggregates in which the heart cells are numerous should occur very early in the sorting-out process because reduction of surface free energy by this means requires far less rearrangement than a commensurate reduction of the total free energy of the system.

It was found that within 17 hours after the onset of aggregation, when sorting out was just beginning to be discernible, a marked depletion of heart cells was evident in the surfaces of the aggregates.[26]

Prediction 2. When a units (heart cells) are very sparse in the population, so that their meeting one another is virtually precluded, minimization of

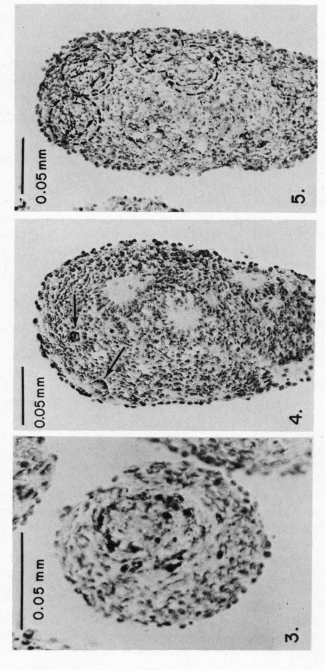

Fig. 3. Section through an aggregate formed by dissociated 5-day heart cells and 7-day retinal cells of chick embryo. Through sorting out, the reconstructed heart tissue has come to be enveloped by the reconstructed retinal tissue. Fig. 4. Section through an aggregate containing 99 % retinal cells and 1 % heart cells. The sparse heart cells, two of which are present in this section, leave the surface but otherwise remain generally distributed. Fig. 5. Section through a heart-retina aggregate early in the process of sorting out. Many small, discrete islets of heart cells (with black inclusions) have formed and are in the process of coalescing. Several islets are encircled on the figure.

surface free energy should cause them to be relegated to subsurface locations within the aggregates. They should be equally stable in all subsurface positions.

It was experimentally established[27] that sorting out of the two types of cells, to yield configurations such as that shown in figure 3, normally was accomplished within $2\frac{1}{2}$ days. Reduction of the proportion of heart cells to 1 % (by volume) of the population yielded aggregates whose surfaces at the end of this time were virtually devoid of heart cells, the latter being otherwise distributed apparently at random within the aggregates (fig. 4). This result, in showing that heart cells do not seek the center, would appear to exclude the possibility that directed migration plays a role in the sorting out of these cells.

Prediction 3. Sorting out should proceed by way of the progressive exchange of heteronomic adhesions for homonomic ones, in the course of which process the potentially internal tissue should appear as a discontinuous phase (i.e., as coalescing islets), whereas the potentially external tissue should constitute a continuous phase.

Histological analysis of heart-retina aggregates fixed after graded intervals in culture bore out prediction 3 (see fig. 5).[26] Similar observations have been reported for the sorting out of mixed amphibian neurula chordamesoderm and endoderm cells[11] and of mixed pigmented retinal and wing bud cells from chick embryos.[28]

Prediction 4. If the distribution of the two phases after segregation is that at which the system is in thermodynamic equilibrium, this same terminal distribution should be approached regardless of the initial distribution of the phases. Thus, lateral fusion of an intact fragment of tissue *b* with an intact fragment of tissue *a* should be followed by the progressive spreading of the one over the surface of the other to yield the same configuration that is ultimately produced through the sorting out of intermixed *a* and *b* cells.

The accuracy of this prediction has been established, to date, for eleven different combinations of tissue fragments and of their dissociated cells.[29,30] In each case, fusion of undissociated fragments of two tissues leads to the progressive envelopment of one fragment by the other, the final disposition of the two tissues being the same as that arrived at when the starting material is a mixed suspension of the corresponding dissociated cells. Of these eleven combinations, nine behaved in the manner described for case 2 (figs. 6 and 7), while two behaved in the manner described for our case 3. The latter showed partial retraction of the earlier continuous, external tissue after segregation within mixed aggregates; correspondingly, they showed only partial enclosure of one fragment by the other after fusion of intact fragments that had never been dissociated.

Prediction 5. In a segregating community composed of two kinds of mutually adhesive, motile units, the less cohesive phase will tend to envelop,

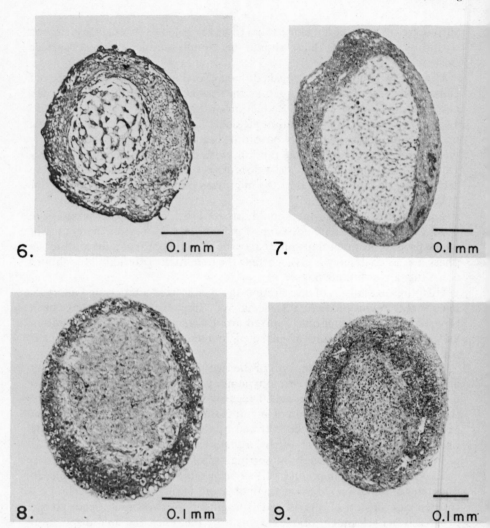

Fig. 6. Section through an aggregate formed by dissociated 4-day limb-bud chondrogenic cells and 5-day heart ventricle cells of chick embryo. The reconstructed heart tissue envelops the now-differentiated cartilage.

Fig. 7. Section through a structure formed by an intact fragment of the chondrogenic zone of a 4-day limb bud laterally fused with a fragment of 5-day heart ventricle. The heart tissue had spread over and enveloped the chondrogenic tissue prior to the deposition of matrix by the latter.

Fig. 8. Section through an aggregate formed by dissociated 5-day heart ventricle cells and 5-day liver cells. The reconstructed liver tissue envelops the reconstructed heart tissue.

Fig. 9. Section through an aggregate formed by dissociated 4-day limb bud chondrogenic cells and 5-day liver cells. The reconstructed liver tissue envelops the chrondrogenic tissue, in which the deposition of matrix has recently begun.

partially or completely, the more cohesive phase at thermodynamic equilibrium. The motile cells of a series of different embryonic tissues constitute a series of phases, each of which is adherent to, yet segregates from, any of the others. Therefore, when the cell populations comprising such a series are intermixed in all possible binary combinations, the mutual positions that they come to assume at equilibrium should establish a hierarchy definable by the specification that if *a* is covered by *b* and *b* is covered by *c*, *a* will be covered by *c*.

In testing this prediction, all possible binary combinations among cell suspensions derived from six different chick-embryo tissues have been used. There are, in all, fifteen different combinations. The segregation patterns obtained do indeed define a hierarchy such that one tissue is reconstituted internally in all combinations, another tissue is reconstituted externally in all combinations, and each remaining tissue falls into a specific intermediate ranking in complete accordance with prediction 5.[30] An example of this behavior is illustrated in figures 6, 8, and 9.

In all respects then, the regroupings of cells in the populations that we have studied proceed along satisfyingly consistent and simple lines. They are precisely what is to be expected on thermodynamic grounds in any system composed of mobile units that are mutually adhesive and among which certain quantitative adhesive relationships exist.

Those of us who have been seeking an explanation for sorting out and tissue reconstruction by dissociated cells have almost unanimously considered it necessary to assume that unlike cells adhere to one another less tenaciously than like cells do [expressed by relation (5)]—a situation requiring selectivity in the mechanism of adhesion. However, the thermodynamic analysis of the situation shows that under the circumstances of relation (5), although sorting out would be expected to occur, the reconstructed tissues would also be expected to continue their mutual self-isolation to a point at which each would come to occupy a portion of the surface of the aggregate (case 3). Moreover, the analysis shows that the most common outcome of segregation in a binary system—the production of a totally internal tissue entirely enveloped by another tissue—fulfills the expectation based upon a set of adhesive relationships quite different from the one that we have previously assumed. In this set of relationships [relation (4)], the unlike cells adhere to one another with a strength intermediate between the strengths of cohesion of the two kinds of like cells. Does logic lead us, then, to postulate the existence of selectivity in the adhesion mechanism itself?

The simplest possible assumption capable of accounting for type-specific differences in the strengths of adhesions between cells is the assumption that adhesive sites of a single kind are scattered more abundantly on the surface of one type of cell than on the surface of another. *W*, the work of adhesion between two cells, would then be directly proportional to the number of adhesive sites that are apposed, per unit of area, at the junction between the two cells. What adhesive relationships would derive from the operation of this simplest of systems?

If the frequency of adhesive sites per unit area on the surfaces of cells a and b is designated f_a and f_b, respectively, the probability of apposition of sites in the cell pairs a–a, b–b, and a–b is given by $(f_a)^2$, $(f_b)^2$, and $(f_a)(f_b)$ for the respective cases. Introducing the proportionality constant k, we may write the equations

$$W_a = k(f_a)^2 \tag{7}$$

$$W_b = k(f_b)^2 \tag{8}$$

$$W_{ab} = k(f_a)(f_b). \tag{9}$$

Following the convention that $W_a \geqslant W_b$, we obtain

$$f_a \geqslant f_b. \tag{10}$$

Multiplying both sides of relation (10) by the value f_a–f_b and rearranging, we obtain

$$\frac{(f_a)^2 + (f_b)^2}{2} \geqslant (f_a)(f_b). \tag{11}$$

Multiplying both sides of relation (10) by the value f_b, we obtain

$$(f_a)(f_b) \geqslant (f_b)^2. \tag{12}$$

Combining relations (11) and (12), we get

$$\frac{(f_a)^2 + (f_b)^2}{2} \geqslant (f_a)(f_b) \geqslant (f_b)^2. \tag{13}$$

Substituting equations (7)–(9) in relation (13), we obtain

$$\frac{W_a + W_b}{2} \geqslant W_{ab} \geqslant W_b. \tag{14}$$

Relation (14), representing the adhesive relationships that would be engendered in this simplest of systems, will be recognized as an expression of the limits represented by relation (4). And the conditions expressed by relation (4) are precisely those that yield, at thermodynamic equilibrium, case 2— the result most commonly obtained experimentally—in which one phase is totally enveloped by the other. It is not necessary to assume the literal existence of discrete adhesive sites as distinguished from nonadhesive sites among which they are distributed. Relation (14) applies to any case in which the force between two mutually adhesive (attractive) bodies is proportional to the product of the individual adhesive (attractive) forces. It may be seen from equations (7)–(9) that the values of W_{ab} generated in this system are given by

$$W_{ab} = \sqrt{W_a} \cdot \sqrt{W_b}. \tag{15}$$

These values are represented by the dotted line in figure 2.

This analysis shows, then, that (1) the mutual sorting out of two kinds of cells to reconstitute tissues, one of which encloses the other, and (2) the spreading of an intact fragment of the one tissue to envelop an intact fragment of the other are precisely the phenomena that are to be expected, in accordance with the principle of minimization of free energy, in the total absence of selectivity in the adhesion mechanism itself. Only quantitative differences in adhesiveness are necessary. The information required in the adhesion mechanism is, in such cases, restricted to "more" and "less".

This does not mean, of course, that molecules of different sorts, on the surfaces either of cells of a given kind or of cells of differing kinds, may not in such cases participate directly in the mediation of adhesions. It merely means that whatever the chemical nature of, or diversity among, the adhesives themselves, the quantitative adhesive relationships among the cells that bear them would be expected to approximate, within the limits shown in figure 2, the relationships derived from the simple postulates that have been outlined. In cases in which, at equilibrium, one tissue covers the other incompletely or not at all, it becomes necessary to assume the additional operation of some other factor or factors, such as an ordered distribution of, or qualitative nonidentity among, adhesive sites.

Morphogenesis and Specificity

Thus we return, at the end, to the beginning: Where is the common denominator? What has sorting out to do with normal morphogenesis? Sorting out, after all, is not known to play a major role in morphogenesis. Such a role, however, *is* played by *spreading*: the spreading of one tissue over the surface of another, or—what is the equivalent—the penetration of one tissue into a mass of another. Differences in cellular adhesiveness that may be built into a system of tissues to bring about the spreading of one tissue over another, or the penetration of one tissue into another, would incidentally (and coincidentally) provide all the conditions required in an artificial mixture of cells for sorting out to occur, and for its morphological result to imitate the anatomy normally produced by mass tissue movements. The foundation for such a thermodynamic analysis, like much of the empirical groundwork upon which it rests, was laid by Holtfreter,[31] whose treatment of the subject has been discussed separately.[32]

Our recognition of the organization everywhere present in the living world has played a prominent role in the development of our biological concepts. It is not surprising that apparent meaningfulness or complexity in the design and functioning of organisms should have led us to assign corresponding attributes to the mechanisms governing the functioning and the design. Yet, as knowledge has grown, complex explanations have had a way of succumbing to relatively simpler ones. Thus, overt vitalism is gone from the scene. Organic molecules, it later developed, could be synthesized by the chemist after all. Proteins were not so simple as to preclude the possibility of their functioning as enzymes; nor was DNA, at a later stage, too simple to provide the vast stores of information for which the proteins, now recognized to be complex, might have seemed a more fitting receptacle.

Although the *adaptedness* brought about through evolution appears complex, the *adaptiveness* that makes evolution possible is born of simplicity. The entire genetic code (and more) is expressible with an alphabet containing only four elements. It would appear that a not inconsiderable amount of the information required to produce, through morphogenetic movement, the anatomy of a body part may be expressed in a code whose sole element is quantity: more versus less. There is, I think, reason to expect that as more realms of biological specificity yield to analysis, their most impressive feature may be the simplicity of the terms in which specificity—information, if you will—can be expressed.

I am indebted to Professor Michael Abercrombie for his penetrating discussions. The original work described here has been supported by grants from the National Science Foundation.

REFERENCES

1. Wilson, H. V. (1907) *J. Exp. Zool.* 5: 245
2. Wilson, H. V. and Penney, J. T. (1930) *J. Exp. Zool.* 56: 73.
3. Huxley, J. S. (1911) *Phil. Trans. Roy. Soc. London* B202: 165; Galtsoff, P. S. (1925) *J. Exp. Zool.* 42: 223; Bronsted, H. V. (1936) *Acta Zool. Stockholm* 17: 75.
4. Harrison, R. G. (1910) *J. Exp. Zool.* 9: 787.
5. Weiss, P. (1941) *Growth* 5, suppl.: 163.
6. Holtfreter, J. (1939) *Arch. Exp. Zellforsch. Gewebezuecht* 23: 169.
7. Herbst, C. (1900) *Arch. Entwicklungsmech. Organ.* 9: 424.
8. Holtfreter, J. (1944) *Rev. Can. Biol.* 3: 220
9. Moscona, A. (1952) *Exp. Cell. Res.* 3: 535.
10. Moscona, A. and Moscona, H. (1952) *J. Anat.* 86: 287; Moscona, A. (1956) *Proc. Soc. Exp. Biol. Med.* 92: 410; (1957) *Proc. Nat. Acad. Sci. U.S.* 43: 184.
11. Trinkaus, J. P. and Groves, P. W. (1955) *Proc. Nat. Acad. Sci. U.S.* 41: 787.
12. Townes, P. L. and Holtfreter, J. (1955) *J. Exp. Zool.* 128: 53.
13. Weiss, P. and Taylor, A. C. (1960) *Proc. Nat. Acad. Sci. U.S.* 46: 1177.
14. Steinberg, M. S. (1958) *Amer. Naturalist* 92: 65.
15. Weiss, P. (1958) *Int. Rev. Cytol.* 7: 391; Weiss, L. (1960) ibid. 9: 187; Weiss, P. (1961) *Exp. Cell. Res.* Suppl. 8: 260; Pethica, B. A. ibid., p. 123; Allen, R. D. ibid., p. 17; Goldacre, R. J. ibid., p. 1; Abercrombie, M. ibid., p. 188; Abercrombie, M. and Ambrose, E. J. (1962) *Cancer Res.* 22: 525.
16. Holtfreter, J. (1939) *Arch. Entwicklungsmech. Organ.* 139: 110.
17. Weiss, P. (1947) *Yale J. Biol. Med.* 19: 235.

18. Tyler, A. (1946) *Proc. Nat. Acad. Sci. U.S.* 32: 195; Weiss, P. (1950) *Quart. Rev. Biol.* 25: 177.
19. DeHaan, R. L. (1958) In *The Chemical Basis of Development*, ed. W. D. McElroy and B. Glass, p. 339. Baltimore: Johns Hopkins Press.
20. Steinberg, M. S. (1962) In *Biological Interactions in Normal and Neoplastic Growth*, ed. M. J. Brennan and W. L. Simpson, p. 127. Boston: Little, Brown.
21. Curtis, A. S. G. (1960) *Amer. Naturalist* 94: 37; (1961) *Exp. Cell Res. Suppl.* 8: 107; (1962) *Biol. Rev. Cambridge Phil. Soc.* 37: 82; (1962) *Nature* 196: 245.
22. Weiss, L. (1962) *J. Theor. Biol.* 2: 236.
23. Stefanelli, A., Zacchei, A. M., and Ceccherini, V. (1961) *Acta. Embryol. Morphol. Exp.* 4: 47.
24. Davies, J. T. and Rideal, E. K. (1961) *Interfacial Phenomena*, pp. 19–23, 34–36. New York: Academic Press.
25. Collins, R. E. and Cooke, Jr., C. E. (1959) *Trans. Faraday Soc.* 55: 1602.
26. Steinberg, M. S. (1962) *Science* 137: 762.
27. Steinberg, M. S. (1962) *Proc. Nat. Acad. Sci. U.S.* 48: 1577.
28. Trinkaus, J. P. (1961) *Colloq. Int. Centre Nat. Rech. Sci. Paris* 101: 209.
29. Steinberg, M. S. (1962) *Proc. Nat. Acad. Sci. U.S.* 48: 1769.
30. Steinberg, M. S. (1970) *J. Exp. Zool.* 173: 395
31. Holtfreter, J. (1943) *J. Exp. Zool.* 94: 261.
32. Steinberg, M. S. (1964) In *Cellular Membranes in Development*, ed. M. Locke, pp. 321–36. New York: Academic Press.

7. Self-Sorting of Isotropic Cells

Narendra Goel, Richard D. Campbell, Richard Gordon,
Robert Rosen, Hugo Martinez, and Martynas Yčas*

Some consequences of a hypothesis proposed by Steinberg on the self-sorting of cells are examined. The hypothesis proposed that two properties, motility and differential adhesion, are sufficient to account for cell sorting. It is assumed that the final configuration reached in pure cell sorting will be one in which the surface free energy of the system will be at a minimum. Immiscible liquid drops are an analog of such a system. We examine the implications of this hypothesis for the morphogenesis of real biological structures. The system is here modeled as a two-dimensional grid whose squares represent cells or ambient medium. A contact edge between the two cells is assigned a λ value that represents the strength or energy of adhesion. This value may be zero or positive. The model is thus a two-dimensional analog of isotropic nondeformable cells. The sum E of the products of number of contact edges and their respective λ values may be regarded as the negative of the total surface free energy of the system. All the configurations (patterns) with maximum E have been found for two and three cell types, together with the constraints on the values of various λs. None of these configurations is histologically significantly interesting. A method for extension to more than three types of cells is also given. The concept of neighboring configurations is introduced. A configuration is neighboring to another if it can be reached from it by cell motility in unit time, assuming certain plausible rules of cell motility. Then certain configurations are at local maxima of E because neighboring configurations have lower E values. Hence, once reached, they may be stable. A number of histologically interesting configurations are shown to be such local maxima. A detailed analysis of a specific case of cell aggregation has also been presented on the basis of our model. A computer simulation of cell movement from a random distribution of two cell types to the maximum E configuration has been made with different and presumably plausible rules of cell motility. None of these trials resulted in the system reaching the maximum E configuration; simulated cells were consistently trapped in configurations of local stability. Thus Steinberg's hypothesis requires special rules of cell motility, which remain to be determined.

1. Introduction

A. Cell Sorting

A wide variety of mechanisms operates during development to produce the final form of an organism. They include cell movement (directed and otherwise), differential rates of cell division and cell death, differential cell–cell

* First published in *J. Theor. Biol.* (1970) 28: 423–68.

adhesions, and other changes in cell properties that may be either autonomous or induced by the presence of another cell type. The multiplicity of morphogenetic mechanisms makes the study of developmental processes difficult; thus, a reasonable way of approaching developmental problems would seem to be to study the various mechanisms in isolated situations. In this paper we confine our attention to cell movements and adhesions that lead to sorting out of cell types.

Since the classical researches of Wilson (1907), it has been known that dissociated sponge cells will reaggregate and reform the whole sponge structure. Initially there was some discussion as to whether these results are due to the sorting out of preexisting cell types or whether the initial disaggregation is followed by a dedifferentiation, random segregation, and then redifferentiation determined by the position of the individual cell. But subsequent work with dissociated embryonic tissues of amphibians, mammals, and birds has clearly demonstrated that the reformation of structures is a true sorting out of preexisting and fixed cell types, at least insofar as cell types can be recognized under the microscope (references in Steinberg, 1964).

B. Steinberg's Hypothesis of Cell Sorting

Steinberg (1963, 1964) was the first to point out explicitly that, in principle, the observed facts on cell self-sorting could be explained on a very simple basis. It suffices to postulate only two cell properties. The first is that cells have what might be called a differential adhesiveness between cells of different types. The second is that cells are motile. Cells with these two properties should then automatically sort themselves out in such a way as to produce structures having maximum stability, or, roughly, the greatest number of cell contacts with the highest degree of mutual adhesiveness.

Steinberg emphasizes that such a final configuration should be independent of the exact physicochemical nature of the adhesive interactions, and of the nature of cell motility. For this reason various types of systems having no close physical similarity will exhibit similar behavior; that is, they will be analogs of one another. Thus one close analog of cell aggregates is a system of mutually immiscible liquids that consists of motile elements, the molecules, having different degrees of affinity for one another.

Steinberg elaborates on his suggestion as follows. One can regard the surface of a cell as having a fixed capacity for binding or adhering to other cells that would then resemble surface free energy. This surface free energy decreases as cell contacts are made. If we have a population consisting of masses of different cell types, the entire population will have a surface free energy that, given cell motility, will tend to minimize. It is then helpful to think of the boundaries of the population of a particular cell type as having a surface free energy to be minimized. In the simplest case in which all the cells are of a single type with more affinity toward one another than toward medium, the cell mass will tend to round up into a sphere, that is, a figure of minimal surface area for a given volume. The analogy with a liquid drop immiscible in ambient liquid is obvious here.

The relations between the cell affinities and the most stable configurations achieved in a system have been investigated in some detail by Steinberg. Using the notation of this paper, the work of cohesion per cell contact in a system of two types of cells, a and b, can be designated λ_{aa}, λ_{ab}, and λ_{bb} between cell types aa, ab, and bb, respectively. If the strength of cohesion between a and b cells is greater than the average of that between a and a and between b and b, or

$$\lambda_{ab} > \frac{\lambda_{aa} + \lambda_{bb}}{2}, \tag{1}$$

the minimum surface energy of the system will be reached when a and b cells alternate, producing in two dimensions a checkerboard pattern that maximizes the area of contact between a and b surfaces. If, on the other hand,

$$\lambda_{ab} < \frac{\lambda_{aa} + \lambda_{bb}}{2}, \tag{2}$$

the condition of minimum free energy will occur when a and b cells are segregated in two homotypic masses. Here, however, there may be three distinct cases. Let

$$\lambda_{aa} > \lambda_{ab} \geqslant \lambda_{bb} \tag{3}$$

so that

$$\frac{\lambda_{aa} + \lambda_{bb}}{2} > \lambda_{ab} \geqslant \lambda_{bb}. \tag{4}$$

a cells, having the maximum cohesiveness, will aggregate and minimize the surface energy of the mass by rounding up into a sphere. b-type cells will aggregate concentrically around the sphere of a-type cells, maximizing b–b contacts and minimizing a–b contacts. This configuration we shall call an onion.

Two other situations exist when

$$\lambda_{aa} \geqslant \lambda_{bb} > \lambda_{ab}. \tag{5}$$

If $\lambda_{ab} = 0$, so that a and b cells do not adhere to their opposite type at all, a and b cells will form two separate spheres. As the value of λ_{ab} rises toward the value λ_{bb}, surface free energy of the system will be minimal when the b sphere partly encloses the sphere of a cells, the area of contact being a function of the value of λ_{ab} relative to λ_{aa} and λ_{bb}. When $\lambda_{ab} > \lambda_{bb}$, enclosure of the a sphere by b becomes complete. It should be noted that surface free energy is minimal whether the two spheres are concentric or not, provided only that enclosure is complete.

The relations are shown in figure 1. They conform completely to configurations taken up by mixtures of immiscible liquids; hence Steinberg refers to his hypothesis as the liquid drop model.

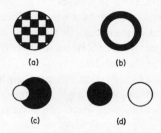

(a) (b)

(c) (d)

Fig. 1. Possible aggregation patterns according to Steinberg (1963, 1964). (a) Checker board, (b) onion, (c) partial enclosure, and (d) separate spheres.

C. Experimental Verification

To demonstrate that cells actually behave in this manner, Steinberg performed a large number of experiments with dissociated embryonic cells, using, for example, unnatural mixtures such as heart-retina, and found that onions were formed. He interpreted this to mean that it is possible to consistently assign λ values to cell types on a linear scale that correctly predicted which cell type, in combination with any other, would form the central mass of the "onion." These results have been fully discussed by Steinberg (1963, 1964) and need not be repeated here. We shall comment on this subject later.

One of Steinberg's further results is, however, of special interest to our problem. Whereas in general randomly dispersed cells with suitable values of λs will aggregate into concentric spheres, in special cases other configurations may be reached. If the number of the more cohesive a cells greatly exceeds the number of the less cohesive b cells, the b cells may become trapped in the interior of the mass of cells and aggregate to form a smaller sphere inside the mass of a cells. This configuration is obviously not one of absolute minimum free energy but is a local minimum, stable only because the b cells, by their motility properties, are likely to never reach the free surface. This situation demonstrates that a configuration may be path dependent, meaning that it may be reachable from some but not from all configurations. The nature of path dependence obviously varies with the rules that govern cell motility.

D. Objectives of This Paper

We investigate here further implications of the interesting hypothesis proposed by Steinberg. Our stepwise approach uses a simplified geometrical–

mathematical model. Initially we make the simplest assumptions; in particular, we assume that cells are nondeformable and isotropic, that is, their adhesive properties are the same in various directions in space. We consider that negative results are as significant as positive ones. If some structures cannot be achieved using minimal postulates, this indicates that other morphogenetic factors must be involved. They may include directional anisotropy of cell adhesions, cell interactions of other than nearest neighbor type, elaboration of cellular matrices, differential cell multiplication and death, change of cell properties with time, and so on. One may then progressively add properties to the model.

The models to be developed are, it is hoped, simple enough to be investigated thoroughly, comprehensive enough to embody in a fairly realistic fashion the basic phenomenological features of the real system, and flexible enough to be extended in a variety of ways as later work indicates.

2. THE BASIC MODEL

For simplicity the models we shall study initially are *two-dimensional*. Let us imagine a region of the plane (possibly infinite) tessellated or subdivided into equal squares. We can index these squares by means of an appropriate coordinate system in the plane. Let us also suppose that we are given fixed numbers $N_1, N_2, \ldots, N_r$ of cells of r types and that each cell is assigned to a square in our grid. The assignment of cells to squares produces a configuration that we shall call a *pattern* (fig. 2). Any square in the grid that does not contain a cell will be assumed to contain *medium*, which if convenient can be treated as another cell type, designated by $r + 1$.

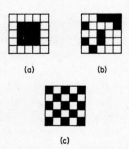

Fig. 2. The model grid and three patterns (a) onion, (b) no preferential pattern, and (c) checkerboard.

To each edge of contact between two cells we assign an affinity, λ, which is a measure of the strength of binding between the cell types. Thus $\lambda_{aa}, \lambda_{ab}$ are the affinities between cells of type a and a, and between a and b, respectively, across the common edge separating them. It is assumed that the affinities between cells are independent of the orientation of the cells. Thus this model is a two-dimensional analog of a system of isotropic cells. We also define an E function, which is the sum of the number of edges multiplied by their λ values in a given pattern. This E can be regarded as the negative or reciprocal of the surface free energy in the sense of Steinberg. The precise relation is not important, because it is only important that as E increases, the surface energy should monotonically decrease. Thus a maximum E will correspond to a minimum surface free energy.

When so desired, the pattern may be allowed to change with time. These changes may be produced by an interchange of cells between certain squares according to rules that define the nature of cell mobility. Cell division and cell death can be simulated by addition or subtraction of cells in the grid. It is important to stress again that the model is purely phenomenological: no specific assumptions are made as to the physical nature of the affinities nor as to the mechanisms of cell motility.

Once the model is defined, it may be used to investigate various properties of cell sorting systems. Here we have studied three.

A. Absolutely Stable (or Maximal) Patterns

Given a finite number of cells, the number of possible patterns is finite. Given a certain choice of values of λ, one or more of them will be configurations with maximum E. One problem, then, is to search for and define such patterns. These patterns, of course, are independent of any rules of cell motility. If all absolutely stable patterns and corresponding constraints on the values of λ are identified, any other pattern will not be an absolutely stable one.

B. Locally Stable Patterns

Realistically, a cell would not be expected to move arbitrarily from one position to any other. It would move initially to a neighboring site, constrained by the presence of other nearby cells. But, given certain motility rules, one configuration could be transformed into another through a series of such elementary cell moves. A configuration reachable from another by a single elementary move will be called a neighbor of the one from which it was derived. Given the concept of neighboring configurations, it is possible to visualize the existence of local maxima of E, or values of E of a configuration that are higher than all values of E of neighboring configurations. Such local maxima make a configuration locally stable because a move away from it entails the passage through configurations of lower E. We discuss here the nature of locally stable patterns.

C. The Path Problem

A further aspect of our work deals with the path problem. It is intuitively obvious that, given some motility rules, not all configurations would be reachable from any arbitrary configuration, because the path from one to another might entail passing through a locally stable configuration. Here we discuss the motility rules that would allow a random pattern of cells to reach an absolutely stable configuration.

3. ABSOLUTELY STABLE PATTERNS

In this section we shall determine absolutely stable patterns, together with the corresponding constraints on λ. For simplicity and to introduce the method and notation, let us first consider the case of two types of cells (which includes the case of one cell type and medium).

A. Patterns Involving Two Types of Cells

Let us denote the types of cells by b and w and let λ_{bb}, λ_{bw}, and λ_{ww} denote the λs for bb, bw (wb), and ww edges, respectively. Let N_b and N_w denote the number of type b and type w cells, respectively.

To determine the absolutely stable patterns, one might think that one would have to generate all the patterns with the same number of cells of the various (here two) types, count the number of edges of various (in this case, three) types, find the E function for each pattern, and thus determine which one is most stable for a given set of values of λ. It turns out that this rather laborious procedure is not necessary if we make use of the fact that because, in our simple model, the cells are nondeformable (or rigid), the number of edges of cells of each type are conserved. In other words, for any configuration,

$$2N_{bb} + N_{bw} = 4N_b \tag{6}$$

$$2N_{ww} + N_{bw} = 4N_w. \tag{7}$$

By definition the E function for a pattern is

$$E = \lambda_{bb}N_{bb} + \lambda_{ww}N_{ww} + \lambda_{bw}N_{bw}. \tag{8}$$

Using equations (6) and (7), equation (8) can be written in either of three equivalent forms:

$$E = N_{bb}\mu_1 + C_1 \tag{9a}$$

$$= -\frac{N_{bw}}{2}\mu_1 + C_2 \tag{9b}$$

$$= N_{ww}\mu_1 + C_3, \tag{9c}$$

where

$$\mu_1 = \lambda_{bb} + \lambda_{ww} - 2\lambda_{bw} \tag{10}$$

$$C_1 = 2\lambda_{ww}(N_w - N_b) + 4\lambda_{bw}N_b \tag{11a}$$

$$C_2 = 2\lambda_{bb}N_b + 2\lambda_{ww}N_w \tag{11b}$$

$$C_3 = 2\lambda_{bb}(N_b - N_w) + 4\lambda_{bw}N_w. \tag{11c}$$

In equations (9), C_1, C_2, and C_3 are constants dependent on λs and the number of cells, but pattern independent. Therefore, for a given set of three λs, that is, given μ_1, equations (9) imply that it is enough to count the number of edges of only one type (say bb) for the purpose of determining the stability of a particular pattern. In writing the conservation relations (6) and (7), we need not consider the boundary of the tessellation for the following reasons. The cells just external to the boundary *must* remain constant (otherwise no solution is possible). We set the boundary sufficiently distant to include all movable cells. Further, we consider the boundary to be made entirely of the predominant cell type because this assumption leads to the most general mathematical expressions. Thus, the system of r cell types may be viewed as cells of $(r - 1)$ cell types embedded in an infinitely large mass of the predominant cell type. (The elimination of boundary considerations is formally justified in the Appendix.)

If one is not aware of the simplification made by the conservation relations (6) and (7), one may say that, in general, one can have many absolutely stable patterns corresponding to the values we give to three λs. But, from equations (9), we see that we can have at most three geometrically (or topologically) different absolutely stable patterns depending upon whether $\mu_1 \gtreqless 0$. Thus, effectively, there is a reduction of three λs (or energy parameters) to one λ (here μ_1). In general, for r types of cells (including medium), there are $r(r + 1)/2$ different kinds of edges and hence λs, and there are r conservation relations. Thus the number of effective λs, that is,

$$\text{no. of } \mu s = \frac{r(r + 1)}{2} - r = \frac{r(r - 1)}{2}. \tag{12}$$

We may now list the possible patterns of absolute stability involving two cell types.

Onion pattern. For $\mu_1 > 0$ from equations (9), the absolutely stable pattern is the one for which N_{bb} and N_{ww} are maximum and N_{bw} is minimum, that is, the patterns in which cells of one type are closely packed and surrounded by cells of other types. Thus for $\mu_1 > 0$, the absolutely stable structure is the onion structure previously described. If w are the predominant cells, b cells will be closely packed, and if b are predominant, w cells will be closely packed. It may be noted that in either case, the minimum value of N_{bw} is not zero.

No preferred pattern. For $\mu_1 = 0$, E is constant and, therefore, all structures are equally stable. Note that this result is more general than the trivial one in which all the λs are equal to one another.

Checkerboard pattern. For $\mu_1 < 0$ from equations (9), the absolutely stable pattern is the one for which N_{bb} and N_{ww} are minimum and N_{bw} is maximum. Such a pattern involves the minority cell dispersed in the predominant cell type; in the case where the minority cells are as close together as possible, the pattern resembles a checkerboard. The three possible patterns are illustrated in figure 2.

Thus, there are only two abosolutely stable patterns for two types of cells: the onion type and the checkerboard type, with the condition that the cells at the boundary are all of one type.

We emphasize that our procedure is valid for nondeformable cells of any shape and, also, in any dimension. To generalize the procedure for any dimension or shape, the only change we have to make is that in the conservation equations (6) and (7), the coefficient 4 on the right-hand side has to be replaced by the number of edges through which a cell can make contact with another cell.

B. Patterns Involving Three types of Cells

We now derive the absolutely stable patterns for a system with three types of cells (or two types of cell and medium). Let us denote the three types by b, w, and e and the number of cells of these types by N_b, N_w, and N_e, respectively. If N_{bb}, N_{bw}, N_{be}, N_{we}, N_{ww}, and N_{ee}, respectively, denote the number of bb, bw, be, we, ww, and ee edges for a pattern, the conservation of edges implies

$$2N_{bb} + N_{bw} + N_{be} = 4N_b \tag{13}$$

$$2N_{ww} + N_{bw} + N_{we} = 4N_w \tag{14}$$

$$2N_{ee} + N_{be} + N_{we} = 4N_e. \tag{15}$$

Here, as for two types of cells, we need not consider edge effects if one cell type (here taken to be e) is predominant. The E function for a pattern is

$$E = \lambda_{bb}N_{bb} + \lambda_{bw}N_{bw} + \lambda_{ww}N_{ww} + \lambda_{be}N_{be} + \lambda_{we}N_{we} + \lambda_{ee}N_{ee}, \tag{16}$$

where λ_{bb} denotes the λ function for contact between b and b cell types, and so forth. Using equations (13) to (15), the right-hand side of equation (16) can be expressed in terms of only three types of edges.

There are $6!/3!3! = 20$ possible distinct groups of three edges. Out of these 20 groups there are three groups, namely those involving

$$N_{bb}, \quad N_{bw} \quad \text{and} \quad N_{be}$$

$$N_{ww}, \quad N_{bw} \quad \text{and} \quad N_{we}$$

$$N_{ee}, \quad N_{be} \quad \text{and} \quad N_{we},$$

in terms of which E cannot be expressed [as can be seen from equations (13) to (15)]. Thus there are 17 possible expressions for E, each involving three types of edges. For the purpose of illustration, let us write E in terms of N_{bb}, N_{bw}, and N_{ww}. From equations (13) and (14),

$$N_{be} = 4N_b - 2N_{bb} - N_{bw} \tag{17a}$$

$$N_{we} = 4N_w - 2N_{ww} - N_{bw}. \tag{17b}$$

Substituting equations (17a) and (17b) into equation (15), we get

$$N_{ee} = 2(N_e - N_b - N_w) + N_{bw} + N_{bb} + N_{ww}. \tag{17c}$$

Substituting equations (17) into equation (16), we obtain

$$\begin{aligned} E = {} & N_{bb}(\lambda_{bb} - 2\lambda_{be} + \lambda_{ee}) + N_{bw}(\lambda_{bw} - \lambda_{be} - \lambda_{we} + \lambda_{ee}) \\ & + N_{ww}(\lambda_{ww} - 2\lambda_{we} + \lambda_{ee}) \\ & + 2[N_b(2\lambda_{be} - \lambda_{ee}) + N_w(2\lambda_{we} - \lambda_{ee}) + N_e\lambda_{ee}]. \end{aligned} \tag{18}$$

The term in square brackets in this equation is a constant (i.e., independent of the pattern).

In a similar fashion we can write the other 16 equivalent forms of E. In column 6 of table 1, we have given explicitly all the 17 forms but have left out the unimportant constant terms. It may be noted that because of the symmetry of the function E for interchange of subscripts b, w, and e, some of the 17 forms can be obtained from other forms by simple interchange of $b \rightleftarrows w$ or $b \rightleftarrows e$ or $w \rightleftarrows e$. If

$$\mu_1 = \lambda_{bb} - 2\lambda_{bw} + \lambda_{ww} \tag{19a}$$

$$\mu_2 = \lambda_{bb} - 2\lambda_{be} + \lambda_{ee} \tag{19b}$$

$$\mu_3 = \lambda_{ww} - 2\lambda_{we} + \lambda_{ee}, \tag{19c}$$

so that

$$\tfrac{1}{2}(\mu_1 + \mu_2 + \mu_3) = \lambda_{bb} + \lambda_{ww} + \lambda_{ee} - \lambda_{bw} - \lambda_{be} - \lambda_{we}, \tag{19d}$$

then all the coefficients of N_{xy} $(x, y = b, w, e)$ in the 17 equivalent forms of E can be expressed as a linear combination of these three μ_i's $(i = 1, 2, 3)$. This proves the statement, made for the case of two types of cells that, effectively, the number of λs is decreased by the number of types of cells in the system.

We now consider the absolutely stable patterns for various values of λ. If

$$\mu_1 = \mu_2 = \mu_3 = 0,$$

then, as can be seen by using any of the 17 equivalent forms, all possible patterns are equally stable.

Table 1. Absolutely stable patterns for the system with three types of cells b, w and e

	Form	N_{bb}	N_{bw}	N_{ww}	N_{be}	N_{we}	N_{ee}	Pattern	Allowed μ's	E, Function form
†	1	1	0	0	0	0	0	b, wxe	$\mu_2 \leq 0,\ \mu_3 \leq 0,\ \mu_2+\mu_3 \leq \mu_1$	
†	1	1	0	0	0	0	0	ⓦ wxe	$\mu_2 > 0,\ \mu_3 \leq 0,\ \mu_2+\mu_3 \leq \mu_1$	
†	1	1	0	0	1	0	0	ⓦ bxe	$\mu_2 \leq 0,\ \mu_3 > 0,\ \mu_2+\mu_3 \leq \mu_1$	$N_{bb}\mu_2 + N_{bw}(\mu_2+\mu_3-\mu_1)/2 + N_{ww}\mu_3$
†	1	1	0	1	0	0	0	bxw	$\mu_2 \leq 0,\ \mu_3 \leq 0,\ \mu_2+\mu_3 > \mu_1$	
	1	1	1	0	1	0	0	ⓑ wxe	$\mu_2 > 0,\ \mu_3 > 0,\ \mu_2+\mu_3 \leq \mu_1$	
	2	2	0	0	0	1	0	b, wxe	$\mu_2 \leq 0,\ \mu_2 \leq \mu_1,\ \mu_3 < 0$	$N_{bb}\mu_2 + N_{bw}(\mu_2-\mu_1)/2 - N_{we}\mu_3/2$
	2	2	0	1	0	0	0	bxw, no we	$\mu_2 \leq 0,\ \mu_2 > \mu_1,\ \mu_3 \geq 0$	
	2	2	1	0	0	1	0	ⓔ wxe	$\mu_2 > 0,\ \mu_1 \geq \mu_2,\ \mu_3 < 0$	
†	3	3	0	0	0	0	0	b, bxw	$\mu_2 \leq \mu_3,\ \mu_1+\mu_3 \geq \mu_2,\ \mu_3 \leq 0$	$N_{bb}(\mu_2-\mu_3) + N_{bw}(\mu_2-\mu_1-\mu_3)/2 + N_{ee}\mu_3$
	3	3	0	1	0	0	1	bxw	$\mu_2 \leq \mu_3,\ \mu_1+\mu_3 < \mu_2,\ \mu_3 > 0$	
†	4	4	0	0	0	0	0	bxw, no be	$\mu_1 \leq \mu_3,\ \mu_2+\mu_3 > \mu_1,\ \mu_3 \leq 0$	$N_{bb}(\mu_1-\mu_3) + N_{be}(\mu_1-\mu_2-\mu_3)/2 + N_{ww}\mu_3$
	4	4	0	0	1	0	1	b, wxe	$\mu_1 \leq \mu_3,\ \mu_2+\mu_3 < \mu_1,\ \mu_3 \leq 0$	
	4	4	0	1	1	0	1	ⓦ bxe	$\mu_1 \leq \mu_3,\ \mu_2+\mu_3 < \mu_1,\ \mu_3 > 0$	
†	5	5	0	0	1	1	1	b, wxe	$\mu_1 < 0,\ \mu_1 > \mu_2,\ \mu_3 < 0$	$N_{bb}\mu_1 + N_{be}(\mu_1-\mu_2)/2 - N_{we}\mu_3/2$
†	6	6	0	0	0	0	1	bxw, no be	$\mu_1 \leq 0,\ \mu_1+\mu_3 \leq \mu_2,\ \mu_3 > 0$	$N_{bb}\mu_1 + N_{be}(\mu_1+\mu_3-\mu_2)/2 + N_{ee}\mu_3$
	6	6	0	0	1	0	0	b, wxe	$\mu_1 \leq 0,\ \mu_1+\mu_3 > \mu_2,\ \mu_3 \leq 0$	
†	6	6	1	0	0	0	1	ⓑ	$\mu_1 > 0,\ \mu_1+\mu_3 \leq \mu_2,\ \mu_3 > 0$	
†	7	7	0	0	0	0	0	bxw, no we	$\mu_2 \leq 0,\ \mu_1 \leq \mu_2,\ \mu_2+\mu_3 \geq \mu_1$	$N_{bb}\mu_2 + N_{ww}(\mu_1-\mu_2) + N_{we}(\mu_1-\mu_2-\mu_3)/2$
	7	7	0	0	1	1	0	b, wxe	$\mu_2 \leq 0,\ \mu_1 \leq \mu_2,\ \mu_2+\mu_3 < \mu_1$	
	7	7	1	0	1	0	0	ⓔ wxe	$\mu_2 > 0,\ \mu_1 \leq \mu_2,\ \mu_2+\mu_3 < \mu_1$	
	8	8	0	0	0	0	0	b, wxe	$\mu_1+\mu_2 \leq \mu_3,\ \mu_1+\mu_3 \leq \mu_2,\ \mu_2+\mu_3 \leq \mu_1$	$N_{bb}(\mu_1+\mu_2-\mu_3)/2 + N_{ww}(\mu_1+\mu_3-\mu_2)/2 + N_{ee}(\mu_2+\mu_3-\mu_1)/2$
	8	8	0	0	0	0	1	bxw	$\mu_1+\mu_2 \leq \mu_3,\ \mu_1+\mu_3 \leq \mu_2,\ \mu_2+\mu_3 > \mu_1$	
†	9	9	0	0	0	0	1	bxw, no we	$\mu_1 \leq 0,\ \mu_1+\mu_3 \geq \mu_2,\ \mu_2 > \mu_1$	$N_{bb}\mu_1 + N_{we}(\mu_2-\mu_1-\mu_3)/2 + N_{ee}(\mu_2-\mu_1)$
	9	9	0	0	0	1	0	b, wxe	$\mu_1 \leq 0,\ \mu_1+\mu_3 < \mu_2,\ \mu_2 \leq \mu_1$	

#	N_{bb} bits	Pattern	Region	E-function
10	0 0 1	b, wxe	$\mu_3 \leq \mu_1,\ \mu_2 < 0,\ \mu_3 \leq 0$	$N_{bw}(\mu_3-\mu_1)/2 - N_{be}\mu_2/2 + N_{ww}\mu_3$
10	0 1 0	bxw, no be	$\mu_3 > \mu_1,\ \mu_2 \geq 0,\ \mu_3 \leq 0$	
10	0 1 1	$(\cdot)\ bxe$	$\mu_3 \leq \mu_1,\ \mu_2 < 0,\ \mu_3 > 0$	$-N_{bw}\mu_2/2 - N_{be}\mu_3/2 - N_{we}\mu_3/2$
11	1 0 1 1	b, wxe	$\mu_1 \geq 0,\ \mu_2 < 0,\ \mu_3 < 0$	
12	0 1	b, wxe	$\mu_1 \geq 0,\ \mu_3 > \mu_2,\ \mu_3 \leq 0$	$-N_{bw}\mu_1/2 + N_{be}(\mu_3-\mu_2)/2 + N_{ee}\mu_3$
12	1 0	bxw, no be	$\mu_1 < 0,\ \mu_2 \geq \mu_3,\ \mu_3 > 0$	
13	0 0	b, wxe	$\mu_1+\mu_2 \geq \mu_3,\ \mu_3 \leq \mu_2,\ \mu_2 \leq 0$	$N_{bw}(\mu_3-\mu_1-\mu_2)/2 + N_{ww}(\mu_3-\mu_2) + N_{ee}\mu_2$
13	1 0	bxw	$\mu_1+\mu_2 < \mu_3,\ \mu_3 \leq \mu_2,\ \mu_2 > 0$	
14	0 0	b, wxe	$\mu_1 \geq 0,\ \mu_2 > \mu_3,\ \mu_2 \leq 0$	$-N_{bw}\mu_1/2 + N_{we}(\mu_2-\mu_3)/2 + N_{ee}\mu_2$
14	1 0	bxw, no we	$\mu_1 < 0,\ \mu_2 \leq \mu_3,\ \mu_2 > 0$	$-N_{be}\mu_2/2 + N_{ww}\mu_1 + N_{we}(\mu_1-\mu_3)/2$
15	0 1	b, wxe	$\mu_2 < 0,\ \mu_1 < 0,\ \mu_1 > \mu_3$	
16 †	0 0 1	b, w no be	$\mu_1+\mu_2 \geq \mu_3,\ \mu_1 < 0,\ \mu_3 > \mu_1$	$N_{be}(\mu_3-\mu_1-\mu_2)/2 + N_{ww}\mu_1 + N_{we}(\mu_3-\mu_1)$
16	0 1 0	b, wxe	$\mu_1+\mu_2 < \mu_3,\ \mu_1 < 0,\ \mu_1 > \mu_3$	
17 †	0 0 1	b, w no we	$\mu_1 \leq 0,\ \mu_1+\mu_2 \leq \mu_3,\ \mu_2 > 0$	
17	0 1 0	b, wxe	$\mu_1 \leq 0,\ \mu_1+\mu_2 > \mu_3,\ \mu_2 \leq 0$	$N_{ww}\mu_1 + N_{we}(\mu_1+\mu_2-\mu_3)/2 + N_{ee}\mu_2$
17 †	1 0 1		$\mu_1 > 0,\ \mu_1+\mu_2 \leq \mu_3,\ \mu_2 > 0$	
†	1 1	(diagram)		
†	1 1	(diagram)	$\mu_2 > \mu_3,\ \mu_1+\mu_3 > \mu_2,\ \mu_2+\mu_3 > \mu_1$	
†	1 1	(diagram)	$\mu_2 < \mu_3,\ \mu_1+\mu_2 > \mu_3,\ \mu_2+\mu_3 > \mu_1$	
†	1 1	(diagram) wxe	$\mu_1 > \mu_3,\ \mu_2 < 0,\ \mu_2+\mu_3 > \mu_1$	
†	1 1	(diagram) bxe	$\mu_1 > \mu_2,\ \mu_2 < 0,\ \mu_2+\mu_3 > \mu_1$	

$$\mu_1 \equiv \lambda_{bb} - 2\lambda_{bw} + \lambda_{ww}, \qquad \mu_2 \equiv \lambda_{bb} - 2\lambda_{be} + \lambda_{ee}, \qquad \mu_3 \equiv \lambda_{ww} - 2\lambda_{we} + \lambda_{ee}$$

In column 3, 0 under N_{bb} stands for the minimum value of N_{bb} and 1 for the maximum value, etc. In column 4 are given the absolutely stable patterns. See text for explanation of the notation. In column 5 are given the regions of μ-space for the various patterns. In column 6 are given the forms of E-function used to derive these regions. The regions marked with † in column 1 are used to make the polyhedron of figure 3. Transitive patterns in the last six rows were not derived by any of the seventeen forms of E-function; these were derived semi-intuitively.

On the other hand, if one μ_i is zero and the other two are equal,

$$\mu_1 = \mu_2 > 0, \qquad \mu_3 = 0,$$

then, by using any of the forms, one finds that for these μ_is, the absolutely stable pattern is the one with b cells clustered together and w cells distributed outside the cluster in any form. Thus, there are an infinite number of absolutely stable patterns, all having b cells clustered together. If

$$\mu_1 = \mu_3 > 0, \qquad \mu_2 = 0$$

then any pattern with w cells clustered together is absolutely stable. For the general case, the systematic approach to identify all absolutely stable patterns is to consider each of the 17 forms of E separately and to find for what values of μ_i there exists one or more absolutely stable patterns. To determine all the absolutely stable patterns from one form, we proceed as follows: in each form, E is expressed in terms of the number of edges of three types. Because E is maximum for an absolutely stable pattern, such a pattern must have the number of edges of these three types, either maximum or minimum. If we denote the maximum by 1 and the minimum by 0, we can have eight possibilities for the number of the three types of edges, namely 000, 001, 010, 100, 110, 101, 011, 111. We investigate each of these eight cases and determine whether we can have a pattern with the number of edges of three types belonging to this case. If so, this will be one of the absolutely stable patterns for suitable values of μ_i implied by the form under consideration. In this way we exhaust all eight cases and find all the absolutely stable patterns implied by this form. We repeat this procedure for other forms.

To determine the values of μ_i for a particular absolutely stable pattern, we set the coefficient of the number of edges of a particular type $\leqslant 0$ if this number is minimum and > 0 if it is maximum. This procedure will ensure maximum E and hence the absolute stability of the pattern under consideration.

As an illustration of our procedure, let us consider for E form 6 of table 1, that is,

$$E = N_{bb}\mu_1 + N_{be}(\mu_1 + \mu_3 - \mu_2)/2 + N_{ee}\mu_3 + \text{constant}. \qquad (20)$$

Suppose we require all N_{bb}, N_{be}, and N_{ee} to be maximum. N_{bb} is maximum when all the b cells are clustered together, but this will not allow N_{be} to be maximum because N_{be} is maximum when all the b cells are intermixed with type-e cells. Likewise, maximum N_{ee} is incompatible with maximum N_{be}. Therefore, N_{bb}, N_{be}, and N_{ee} cannot be simultaneously maximum, and hence we will not have an absolutely stable pattern if the coefficients of N_{bb}, N_{be}, and N_{ee} in equation (20) are positive. On the other hand, if we require both N_{bb} and N_{ee} to be maximum and N_{be} to be minimum, we do have a pattern with these properties: namely, all b cells clustered, this cluster covered by

w cells tightly packed, and this cluster of b and w cells surrounded by e cells. This is an onion pattern with b inside, followed by w and e. Such a pattern will be absolutely stable if the coefficients of N_{bb} and N_{ee} in equation (20) are positive and that of $N_{be} \leqslant 0$, that is,

$$\mu_1 > 0, \qquad \mu_1 + \mu_3 \leqslant \mu_2, \qquad \mu_3 > 0.$$

We repeat this procedure for all 17 forms. The results are summarized in table 1. In the second column we have assigned a number to the E-function form listed in column 6. In the third column we have given the values (maximum or minimum) for number of various types of edges used to obtain the absolutely stable pattern given in column 4. In column 3, 0 stands for minimum and 1 stands for maximum. The values of μ_i for these patterns are given in column 5. In column 4, $\text{⬭}(b)w$ stands for onion pattern with b cells inside surrounded by w cells, which, in turn, are surrounded by e cells; b, wxe stands for complete dispersion of b and w cells with e cells; $\text{⬭}(b)$ wxe stands for all b cells clustered together and w and e cells completely dispersed; bxw stands for checkerboard pattern between b and w cells surrounded by e cells; bxw no be stands for checkerboard pattern between b and w cells such that there is no contact between b and e cells; $\text{⬭}(w)b$, $\text{⬭}(w)$ bxe, and bxw no we have similar meanings. We may remark that bxw, bxw no be and bxw no we patterns depend upon the number of cells. For bxw, $N_b \simeq N_w$, and for bxw no we, $N_w \lesssim N_b$.

Our systematic procedure determines all the absolutely stable patterns for three types of cells. One may ask the question: Do the values of μ_i in column 4 cover the whole μ_1, μ_2, μ_3 space? The answer is no. There are three regions defined by (a) $\mu_1 < \mu_2 + \mu_3, \mu_2 < \mu_1 + \mu_3$, and $\mu_3 < \mu_1 + \mu_2$; (b) $\mu_3 < \mu_1, \mu_3 < 0$ and $\mu_1 < \mu_2 + \mu_3$; and (c) $\mu_2 < \mu_1, \mu_2 < 0$ and $\mu_1 < \mu_2 + \mu_3$, in which the absolutely stable patterns cannot be determined in this way.

These three regions can be determined either by a geometric or an algebraic method. We will here give the geometric method and will discuss the algebraic method in the next paper (Goel and Leith, 1970). In the geometric method we construct a polyhedron in μ_1, μ_2, μ_3 space so that its various faces define the regions giving various absolutely stable structure. Such a polyhedron can be constructed by using figure 3 and following the instructions given with it. The patterns drawn are listed in table 1. If μ_1, μ_2, and μ_3 are such that the vector defined by them passes through a region with a particular pattern on it, this will be the absolutely stable pattern for these μ values. This geometrical method will fail for systems with four or more types of cells because there one will deal with a six or more dimensional μ space.

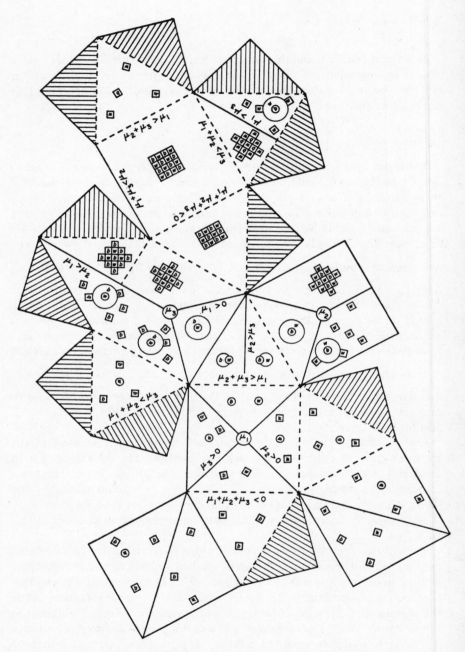

Fig. 3. Polyhedron in μ-space defining regions giving maximally stable structures; μ_1, μ_2, μ_3 axes intersect the surface in the center of the square faces. Intersections of the three positive axes are marked in circles. Instructions for assembling are: (a) Cut along outer lines only; (b) fold back along each dotted line; (c) hatched triangles are flaps. With tape or glue, fasten these to the undersurfaces of the square faces.

114

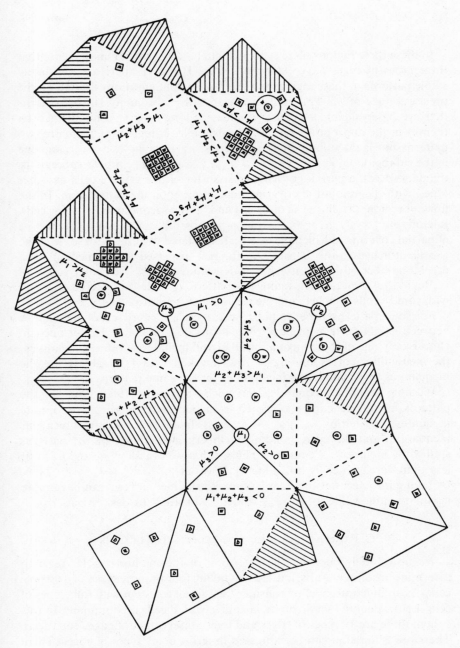

(Cut out and construct according to legend to fig. 3.)

In these three regions there is no absolutely stable pattern in the sense that three types of edges have extremum value. However, there are absolutely stable patterns in these regions but their exact configurations depend on the precise values of μ_i. This is contrary to other absolutely stable patterns that are independent of precise μ values as long as the vector defined by them is in the appropriate region. For example, in region (a), there are two patterns, one in the subregion $\mu_2 > \mu_3, \mu_2 < \mu_1 + \mu_3, \mu_1 < \mu_2 + \mu_3$ and one in the subregion $\mu_2 < \mu_3, \mu_3 < \mu_1 + \mu_2, \mu_1 < \mu_2 + \mu_3$. In both subregions, the pattern is the partial covering of a cluster of one type of cells by other types, with the amount of covering depending upon the values of μ_i. In the first subregion, the cluster is of b cells and, in the second one, it is of w cells. Region (b) $\mu_1 > \mu_3, \mu_3 < 0, \mu_2 + \mu_3 > \mu_1$ possibly contains a different type of partial covering in which the b cells are clustered together with some of the w cells attached to this cluster and the rest dispersed in the medium. The precise number of w cells attached is determined by the particular values of μ. Region (c) contains an analogous pattern of w cells clustered together with some b cells attached and the remaining b cells dispersed.

For some patterns, for example, b, wxe, there are many sets of inequalities between μs obtained by using various forms of E (see table 1). But one can choose only one of these inequalities that will include the other inequalities; the inequalities corresponding to rows marked with † in table 1 are the inequalities that include the inequalities for the same patterns.

We note that the inequalities derived by Steinberg (1964, 1965) for the patterns he studied can be obtained from our corresponding appropriate inequalities by letting $\lambda_{be} = \lambda_{ee} = \lambda_{we} = 0$. In table 2 we have given our inequalities and the corresponding Steinberg inequalities for the patterns studied by Steinberg. We must emphasize that we can allow μ_2 and μ_3 to be less than zero, which is not allowed if we let $\lambda_{be} = \lambda_{ee} = \lambda_{we} = 0$. Thus, by taking into account the interaction with the medium, we can have more flexibility in the values of $\lambda_{bb}, \lambda_{ww}$, and λ_{bw} and still get a desired pattern.

C. Patterns with Four and More Types of Cells

We now briefly consider the extension of the methods so far used to determine absolutely stable patterns involving four or more types of isotropic cells. As an illustration let us consider the case of a system with four types of cells. Let us denote them by b, w, m, and e, with e cells predominant. In this system there are 10 types of edges and four conservation of edges equations. Therefore E function can be expressed in terms of six types of edges. There are, altogether, $10!/6!6! = 210$ groups of types of edges, each group consisting of six types of edges. Therefore it is convenient to write a FORTRAN program for a computer to solve the four conservation equations in terms of various groups of edges. It turns out that out of 210 groups, 69 groups are

Table 2. Absolutely stable patterns discussed by Steinberg and the corresponding regions of μ-space

Pattern	Region of μ-space	Region of λ-space with $\lambda_{be} = \lambda_{we} = \lambda_{ee} = 0$, or, Steinberg's inequalities
⊚	$\mu_1 > 0$, $\mu_2 \geqslant \mu_1 + \mu_3$, $\mu_3 > 0$	$\lambda_{bb} - 2\lambda_{bw} + \lambda_{ww} > 0$, $\lambda_{bw} \geqslant \lambda_{ww}$, $\lambda_{ww} > 0$
⊚	$\mu_1 > 0$, $\mu_3 \geqslant \mu_1 + \mu_2$, $\mu_2 > 0$	$\lambda_{bb} - 2\lambda_{bw} + \lambda_{ww} > 0$, $\lambda_{bw} \geqslant \lambda_{bb}$, $\lambda_{bb} > 0$
⊙ ⊙	$\mu_2 > 0$, $\mu_3 > 0$, $\mu_2 + \mu_3 \leqslant \mu_1$	$\lambda_{bb} > 0$, $\lambda_{ww} > 0$, $\lambda_{bw} \leqslant 0$ i.e., $\lambda_{bw} = 0$
◑ ⋅	$\mu_2 > \mu_3$, $\mu_2 < \mu_1 + \mu_3$, $\mu_1 < \mu_2 + \mu_3$	$\lambda_{bb} > \lambda_{ww}$, $\lambda_{ww} > \lambda_{bw}$, $\lambda_{bw} > 0$
◑ ⋅	$\mu_2 < \mu_3$, $\mu_3 < \mu_1 + \mu_2$, $\mu_1 < \mu_2 + \mu_3$	$\lambda_{bb} < \lambda_{ww}$, $\lambda_{bb} > \lambda_{bw}$, $\lambda_{bw} > 0$
bxw	$\mu_1 + \mu_2 \leqslant \mu_3$, $\mu_1 + \mu_3 \leqslant \mu_2$, $\mu_2 + \mu_3 > \mu_1$	$\lambda_{bw} \geqslant \lambda_{bb}$, $\lambda_{bw} \geqslant \lambda_{ww}$, $\lambda_{bw} > 0$

to be discarded for the same reasons we discarded three out of 20 groups for the case of three types of cells (Goel and Leith, 1970). One has, therefore, to consider $210 - 69 = 141$ forms of E, each involving six types of edges.

From the extension of the discussion in the previous section, it follows that an absolutely stable pattern will result if each of the number of any six types of edges is either maximal or minimal. If a pattern does satisfy these conditions, it will be absolutely stable provided the coefficients of the number of six types of edges are >0 or $\leqslant 0$, depending upon whether the number of the corresponding type of edge is maximum or minimum.

For each form there are $2^6 = 64$ possibilities for the number (maximum or minimum) of edges. One has, therefore, to investigate $141 \times 64 = 9024$ cases for possible absolutely stable patterns. Once again one can seek the help of a computer. It is possible to write a FORTRAN program (Goel and Leith, 1970) that will identify most of the cases for which there is no absolutely stable structure and determine the absolutely stable structures, together with the constraints on the values of λ.

However, without going through this detailed systematic procedure, we can find some of the absolutely stable patterns together with the constraints on the values of six μs. These patterns are listed in table 3, where we have also included the types of edges that have either maximum (1) or minimum values (0). We have not included the patterns that can be obtained by interchanging b with w or with m. For these patterns the constraints on μs can

Table 3. Some of the absolutely stable patterns for a system with four types of cells

Pattern	Types of edges									
	N_{bb}	N_{bw}	N_{bm}	N_{be}	N_{ww}	N_{wm}	N_{we}	N_{mm}	N_{me}	N_{ee}
ⓑ ⓦ ⓜ	1	0	0		1	0		1		
b, w, mxe	0	0	0	1	0	0	1	0	1	0
ⓑ w, mxe	1	0	0		0	0	1	0	1	
(bxw) mxe	0	1	0		0	0		0	1	
((bxw)m) e	0	1			0	0		0		1

For these particular patterns the constraints on μ's can be obtained using the methods applicable to a system with three types of cells. Notation as in Table 1.

be obtained by the same methods as used for a system with three types of cells. One can easily check that the method of this section is applicable for patterns of the type listed in table 3 systems with any number of cell types.

However, for a pattern that immediately comes to the mind, namely, the onion pattern, the methods so far used fail. For example, for the onion pattern

N_{bb} and N_{ee} are maximum and N_{bm}, N_{be}, and N_{we} are minimum. The numbers of other types of edges do not have a maximum or minimum value for this pattern. Thus, the number of types of edges whose numbers have either a maximum or minimum value falls short by one, the minimum number necessary for the application of the methods so far used.

There is an indirect but not completely satisfactory procedure to determine the conditions for the absolute stability of the above onion pattern. This method consists of treating w and m cells as identical and finding the conditions for the absolute stability of pattern ⓑ w e, and then treating b and w cells as identical and finding the conditions for the absolute stability of the pattern ⓑ^{m}ᵉ. But one can also use a direct and satisfactory alternate procedure that will now be described.

This procedure, which we call the method of replacement, involves removing some of the cells from the inside of an absolutely stable (or maximal) onion pattern, replacing them by another type of cells, and finding the conditions under which the new onion pattern is absolutely stable. In other words, if we know that an onion pattern is stable for a system with n types of cells, we can find the conditions on λs under which the onion with $n + 1$ types of cells will be absolutely stable. This method demonstrates the absolute stability of all the onion patterns with two or more layers, each made up of different types of cells (including the case where the innermost clump of the onion is a checkerboard between two types of cells), and, as follows from the above discussion, the method of the previous section demonstrates the absolute stability of other patterns for a system with any number of cells. Thus, using these two methods, presumably we can determine all the absolutely stable patterns and the corresponding conditions on λs for a system with any number of types of cells.

D. Generation of Maximal Patterns by Method of Replacement

Before we derive the conditions on λs under which an onion with $n + 1$ types of cells is absolutely stable, provided the onion with n types of cells is absolutely stable, we give a preliminary result that demonstrates the kind of reasoning we use. Let $C(N_b, N_w, N_e)$ be the class of patterns each of which has N_b cells of type b, N_w cells of type w, and N_e cells of type e (the medium cells).

Let $E(w \to b)$ denote the change in E function of a given pattern when we replace a w cell by a b cell. Given the pattern Q maximal in $C(N_b - 1, N_w + 1, N_e)$, let E_0 be the maximum value of $E(w \to b)$ in Q and choose a $w \to b$ replacement for which $E(w \to b) = E_0$. The resulting pattern is then maximal in $C(N_b, N_w, N_e)$ if there is no other member of $C(N_b - 1, N_w + 1, N_e)$ having an $E(w \to b)$ larger than E_0.

This result becomes self-evident when we consider that any member of $C(N_b, N_w, N_e)$ can be obtained from any member of $C(N_b - 1, N_w + 1, N_e)$ by first performing a suitable permutation and then making a $w \to b$ conversion. If Q is maximal in $C(N_b - 1, N_w + 1, N_e)$, then the E function $E(Q)$ of Q satisfies the relation $E(Q) \geqslant E(\pi Q)$ for arbitrary permutation π. Accordingly, the E value of πQ followed by a $w \to b$ conversion must be equal to or less than the E value of Q by its stated $w \to b$ conversion if πQ contains no $w \to b$ conversion for which $E(w \to b) > E_0$. For completeness, we note that a similar result holds if Q is maximal in $C(N_b + 1, N_w - 1, N_e)$ and we choose a $b \to w$ conversion for which $E(b \to w)$ is maximal in Q.

A simple application of this result is to a pattern consisting of only w cells surrounded by medium, that is, Q consisting of a ball of w cells surrounded by medium. If we assume that $\lambda_{ww} > 0$ and $\lambda_{ee} = \lambda_{we} = 0$, such a pattern is easily shown to be maximal from the general E function expression:

$$E(Q) = \lambda_{ww}N_{ww} + \lambda_{bw}N_{bw} + \lambda_{bb}N_{bb} + \lambda_{we}N_{we} + \lambda_{be}N_{be} + \lambda_{ee}N_{ee}. \quad (21)$$

This reduces to

$$E(Q) = \lambda_{ww}N_{ww} \quad (22)$$

and is clearly maximal if all the w cells are in the form of a ball presenting least surface to the medium. Our question is: can we replace a w cell in Q by a b cell and have the resulting pattern maximal? To answer this question it is first necessary to obtain an expression for $E(w \to b)$ for an arbitrary pattern. If $E(w)$ is the contribution of a w cell to the function (which, for convenience, we call the E value of a w cell) and it is in contact with X_w w cells and X_b b cells, then

$$E(w) = \lambda_{ww}X_w + \lambda_{bw}X_b. \quad (23a)$$

Replacing w by a b cell will give an E value to a b as in

$$E(b) = \lambda_{bb}X_b + \lambda_{bw}X_w. \quad (23b)$$

Hence,

$$E(w \to b) = E(b) - E(w) = (\lambda_{bb} - \lambda_{bw})X_b + (\lambda_{bw} - \lambda_{ww})X_w. \quad (24)$$

If we assume that $\lambda_{bb} - \lambda_{bw} > 0$, $\lambda_{bw} - \lambda_{ww} > 0$, and $\lambda_{bb} - \lambda_{bw} > \lambda_{bw} - \lambda_{ww}$, then $E(w \to b)$ would be maximized by choosing a w cell having the largest number of b contacts and the least number of w contacts. Because in Q there are no b cells, $E(w \to b)$ would be maximal by choosing a w cell having as many w contacts as possible, meaning a w cell that is in the interior of the ball. Clearly, no permutation of Q will yield a w cell having more w contacts than this, from which it then follows that replacing an interior w cell by a b cell results in a maximal pattern if the above relations for the λs are satisfied, although for the special case under consideration we only really need $\lambda_{bw} - \lambda_{ww} > 0$.

If, in the resulting pattern obtained above, we replace a w cell that is in contact with the single b cell, and still interior to the ball, by a b cell, the resulting pattern is again maximal by the same reasoning. Unfortunately, the reasoning is no longer applicable when we arrive at a stage such that a permutation will yield a pattern in which a w cell is in contact with only b cells. Thus we cannot, by this reasoning, prove the maximality of an onion structure in which a ball of b cells is surrounded by w cells under the conditions that $\lambda_{bb} - \lambda_{bw} > \lambda_{bw} - \lambda_{ww} > 0$. We now proceed to illustrate the slightly different mode of reasoning that is required.

In the above ball of w cells, suppose we choose a collection of n_w interior cells that offers least contact with its complement. If N_{ww}^i and N_{ww}^e are, respectively, the number of internal and external $w - w$ contacts that this collection possesses, we want the ratio N_{ww}^i/N_{ww}^e to be as large as possible. We would like to show that if this collection is replaced by an equal number of b cells, the resulting pattern is maximal provided $\lambda_{bb} - \lambda_{bw} > \lambda_{bw} - \lambda_{ww} > 0$. Letting $\{n_w\}$ be the collection of w cells removed,

$$E(\{n_w\}) = \lambda_{ww}N_{ww}^i + \lambda_{ww}N_{ww}^e. \tag{25}$$

Also, if $\{n_b\}$ is the collection of replacing b cells,

$$E(\{n_b\}) = \lambda_{bb}N_{ww}^i + \lambda_{bw}N_{ww}^e. \tag{26}$$

Accordingly, the change in E value of the pattern upon making the replacement is given by

$$E(\{n_b\}) - E(\{n_w\}) = (\lambda_{bb} - \lambda_{ww})N_{ww}^i + (\lambda_{bw} - \lambda_{ww})N_{ww}^e. \tag{27}$$

Further, if k is the number of contacts allowed per cell ($k = 4$ for our two-dimensional cells), then $kn_w = 2N_{ww}^i + N_{ww}^e$, and so

$$E(\{n_b\}) - E(\{n_w\}) = (\lambda_{bb} - \lambda_{bw})kn_w/2 + [\lambda_{bw} - \tfrac{1}{2}(\lambda_{bb} + \lambda_{ww})]N_{ww}^e. \tag{28}$$

Next to be considered is an arbitrary permutation of Q followed by an exchange of n_w cells with b cells, just as in the first example. Letting $\{\bar{n}_w\}$ denote the new collection of w cells removed, with $\bar{n}_w = n_w$, then

$$E(\{\bar{n}_w\}) = \lambda_{ww}\bar{N}_{ww}^i + \lambda_{ww}\bar{N}_{ww}^e \tag{29}$$

$$E(\{\bar{n}_b\}) = \lambda_{bb}\bar{N}_{ww}^i + \lambda_{bw}\bar{N}_{ww}^e. \tag{30}$$

Using the relation

$$k\bar{n}_w = 2\bar{N}_{ww}^i + \bar{N}_{ww}^e + \bar{N}_{we}^e, \tag{31}$$

then

$$E(\{\bar{n}_b\}) - E(\{\bar{n}_w\}) = (\lambda_{bb} - \lambda_{bw})kn_w/2 - (\lambda_{bb} - \lambda_{bw})\bar{N}_{we}^e/2$$
$$+ [\lambda_{bw} - \tfrac{1}{2}(\lambda_{bb} + \lambda_{ww})]\bar{N}_{ww}^e. \tag{32}$$

The difference in E in the two exchanges [equations (32) to (28)] is given by

$$(32) - (28) = [\lambda_{bw} - \tfrac{1}{2}(\lambda_{bb} + \lambda_{ww})](\bar{N}_{ww}^e - N_{ww}^e) - \tfrac{1}{2}(\lambda_{bb} - \lambda_{ww})\bar{N}_{we}^e. \tag{33a}$$

Our assumption that the ratio N_{ww}^i/N_{ww}^e of the first exchange be as large as possible and that only interior cells be involved ensures that $\bar{N}_{ww}^e + \bar{N}_{we}^e \geq N_{ww}^e$. Substitution of this inequality into (33a) under the additional assumption that $\lambda_{bw} - \tfrac{1}{2}(\lambda_{bb} + \lambda_{ww}) \leq 0$ yields the inequality

$$(32) - (28) \leq (\lambda_{ww} - \lambda_{bw})\bar{N}_{we}^e, \tag{33b}$$

which is clearly nonpositive if $\lambda_{ww} \leqslant \lambda_{bw}$. This gives a proof of the maximality of the onion pattern

under the constraints $\lambda_{be} = \lambda_{we} = \lambda_{ee} = 0$, $\lambda_{bb} - \lambda_{bw} > \lambda_{bw} - \lambda_{ww} > 0$.

Desired next is a proof of the absolute stability of the n-layered onion structure. We will do this by induction, in the sense that a p-layered onion will be assumed to be maximal, and we will derive from this a maximal $p + 1$-layered onion. For purposes of notation we will assume that the 0-layer is medium and the numbering of the layers is from outside in, so that layer p is the core. Layer j consists of cells of c_j and, as in the previous example, we will replace some of the type c_p cells with cells of type c_{p+1}. The conservation relations needed are

$$kn_p = 2N^i_{p\cdot p} + N^e_{p\cdot p} \tag{34a}$$

and

$$k\bar{n}_p = 2\bar{N}^i_{p\cdot p} + \sum_{j=0}^{p} \bar{N}^e_{p\cdot j}. \tag{34b}$$

As before, it is immediately deduced that

$$E(\{n_p\}) = \lambda_{p\cdot p}N^i_{p\cdot p} + \lambda_{p\cdot p}N^e_{p\cdot p} \tag{35a}$$

$$E(\{n_{p+1}\}) = \lambda_{p+1\cdot p+1}N^i_{p\cdot p} + \lambda_{p+1\cdot p}N^e_{p\cdot p} \tag{35b}$$

and

$$E(\{n_p\}) - E(\{n_{p+1}\}) = \tfrac{1}{2}kn_p(\lambda_{p+1\cdot p+1} - \lambda_{p\cdot p})$$
$$+ [\lambda_{p+1\cdot p} - \tfrac{1}{2}(\lambda_{p+1\cdot p+1} + \lambda_{p\cdot p})]N^e_{p\cdot p} \tag{36}$$

$$E(\{\bar{n}_p\}) = \lambda_{p\cdot p}\bar{N}^i_{p\cdot p} + \sum_{j=0}^{p} \lambda_{p\cdot j}\bar{N}^e_{p\cdot j} \tag{37}$$

$$E(\{\bar{n}_{p+1}\}) = \lambda_{p+1\cdot p+1}\bar{N}^i_{p\cdot p} + \sum_{j=0}^{p} \lambda_{p+1\cdot j}\bar{N}^e_{p\cdot j} \tag{38}$$

$$E(\{\bar{n}_{p+1}\}) - E(\{\bar{n}_p\}) = (\lambda_{p+1\cdot p+1} - \lambda_{p\cdot p})\bar{N}^i_{p\cdot p} + \sum_{j=0}^{p} (\lambda_{p+1\cdot j} - \lambda_{p\cdot j})\bar{N}^e_{p\cdot j}$$

$$= (\lambda_{p+1\cdot p+1} - \lambda_{p\cdot p})\left(\tfrac{1}{2}k\bar{n}_p - \tfrac{1}{2}\sum_{j=0}^{p} \bar{N}^e_{p\cdot j}\right)$$

$$+ \sum_{j=0}^{p} (\lambda_{p+1\cdot j} - \lambda_{p\cdot j})\bar{N}^e_{p\cdot j}$$

$$= (\lambda_{p+1 \cdot p+1} - \lambda_{p \cdot p}) \tfrac{1}{2} k \bar{n}_p$$

$$+ [\lambda_{p+1 \cdot p} - \tfrac{1}{2}(\lambda_{p+1 \cdot p+1} + \lambda_{p \cdot p})] \bar{N}_{p \cdot p}^e$$

$$-\tfrac{1}{2}(\lambda_{p+1 \cdot p+1} - \lambda_{p \cdot p}) \sum_{j=0}^{p-1} \bar{N}_{p \cdot j}^e + \sum_{j=0}^{p-1} (\lambda_{p+1 \cdot j} - \lambda_{p \cdot j}) \bar{N}_{p \cdot j}^e. \tag{39}$$

The difference in exchange energy is thus given by

$$(39) - (36) = [\lambda_{p+1 \cdot p} - \tfrac{1}{2}(\lambda_{p+1 \cdot p+1} + \lambda_{p \cdot p})](\bar{N}_{pp}^e - N_{pp}^e)$$

$$-\tfrac{1}{2}(\lambda_{p+1 \cdot p+1} - \lambda_{p \cdot p}) \sum_{j=0}^{p-1} \bar{N}_{p \cdot j}^e + \sum_{j=0}^{p-1} (\lambda_{p+1 \cdot j} - \lambda_{p \cdot j}) \bar{N}_{p \cdot j}^e. \tag{40a}$$

As in the previous example, choice of the cells replaced ensures that

$$\bar{N}_{p \cdot p}^e + \sum_{j=0}^{p-1} \bar{N}_{p \cdot j}^e \geqslant N_{p \cdot p}^e.$$

Substituting this inequality into equation (40a) under the additional assumption that $\lambda_{p+1 \cdot p} - \tfrac{1}{2}(\lambda_{p+1 \cdot p+1} + \lambda_{p \cdot p}) \leqslant 0$ yields

$$(39) - (36) = (\lambda_{p \cdot p} - \lambda_{p+1 \cdot p}) \sum_{j=0}^{p-1} \bar{N}_{p \cdot j}^e + \sum_{j=0}^{p-1} (\lambda_{p+1 \cdot j} - \lambda_{p \cdot j}) \bar{N}_{p \cdot j}^e, \tag{40b}$$

which is clearly nonpositive if $\lambda_{p+1 \cdot p} > \lambda_{p \cdot p}$ and $\lambda_{p \cdot j} > \lambda_{p+1 \cdot j}$ for $j \neq p$. Demonstrated, therefore, is the maximality of the n-layered onion structure under appropriate conditions.

We conclude this section by making a few comments on the relation between transitive and nontransitive relations between λs and the absolute stability. According to the interpretation of Steinberg, affinities between cells, or the lambda values, differ only quantitatively; there are no specific affinities between the same or, say, between two types of cells only. The lambdas are scalar quantities. From the physical point of view, this of course is not the only situation conceivable. Cells might be imagined to have special types of "hooks," so that they have affinity only for other cells with complementary hooks. The first relation may be called transitive, the second nontransitive. Thus experiment has to determine which situation actually prevails.

Steinberg (1964) states, as a summary of his experimental results, "If tissue a segregates internally to tissue b, and tissue b segregates internally to tissue c, then tissue a will be found to segregate internally to tissue c." This is apparently also intended to be a statement of the results to be expected even if the affinities are nontransitive, but in fact it is not so, because the affinities between cells of types a and b and b and c do not imply anything as to the affinities between a and c. The following explanation shows this relationship.

Let the absolutely stable (maximum E) pattern of b and w cells be b enclosed by w, and similarly that of w and x cells be a pattern of w enclosed by x. Let e denote the medium common to both systems. The question we then ask is whether this implies anything about the maximum E pattern of b and x. The first two observations and the previous discussion imply:

$$\lambda_{bb} - 2\lambda_{bw} + \lambda_{ww} > 0 \quad (41a); \qquad \lambda_{ww} - 2\lambda_{wx} + \lambda_{xx} > 0 \quad (42a)$$

$$\lambda_{we} + \lambda_{bw} \geqslant \lambda_{be} + \lambda_{ww} \quad (41b); \qquad \lambda_{xe} + \lambda_{wx} \geqslant \lambda_{we} + \lambda_{xx} \quad (42b)$$

$$\lambda_{ee} - 2\lambda_{we} + \lambda_{ww} > 0 \quad (41c); \qquad \lambda_{ee} - 2\lambda_{xe} + \lambda_{xx} > 0 \quad (42c).$$

Adding equations (41a) and (42a), we get

$$\lambda_{bb} + \lambda_{xx} + 2\lambda_{ww} > 2\lambda_{bw} + 2\lambda_{wx}. \tag{43}$$

Multiplying equations (41b) and (42b) by 2 and adding both resulting equations to equation (43), we get

$$\lambda_{bb} - \lambda_{xx} > 2(\lambda_{be} - \lambda_{xe}). \tag{44}$$

Adding equation (42c) to equation (44), we finally have

$$\lambda_{bb} + \lambda_{ee} > 2\lambda_{be}. \tag{45}$$

From equations (42)–(45) we conclude that the above-mentioned observations tell us nothing about λ_{bx} and hence do not imply anything as to the absolutely stable pattern between b and x, although in the special case of $\lambda_{be} = \lambda_{xe} = \lambda_{ee} = 0$, equation (44) does imply that $\lambda_{bb} > \lambda_{xx}$. Therefore, the enclosure of x cells by b cells is excluded. If we further assume that $\lambda_{bx} = 0$, equations (42c) and (45) and the discussion earlier in this section imply that b cells and x cells will be completely segregated. On the other hand, if instead of assuming $\lambda_{bx} = 0$, we assume that $\lambda_{xx} \leqslant \lambda_{bx} < (\lambda_{bb} + \lambda_{xx})/2$, then b cells will be aggregated internally to x cells. Thus both transitive and non-transitive relations are compatible with the model. It should be explicitly noted that whether λs are transitive or not, the number of possible absolutely stable patterns will not change. This is because our analysis of stable patterns has covered all the μ-space (which is a complete projection of λ space).

Experimentally most results of cell-sorting experiments, including those of Steinberg, are compatible with transitive relations. But there are a few exceptions. For example, in the experiments of Townes and Holtfreter (1955) on the ectoderm (a), mesoderm (m), and endoderm (b) from an amphibian neurula, both ectoderm and endoderm enclose the mesoderm when these tissues are placed in contact with each other, but ectoderm and endoderm segregate from each other. Thus it may be concluded that $\lambda_{mm} > \lambda_{aa}$, $\lambda_{mm} > \lambda_{bb}$, and $\lambda_{ab} = 0$.

We may also add that our model implies that if tissue a segregates internally to tissue b, tissue b segregates internally to tissue c, and tissue a segregates internally to tissue c: then, if all three tissues a, b, and c are allowed to aggregate, tissue a will be segregated internally to tissue b, which in turn will be segregated internally to tissue *c*. This succession follows from the analysis, similar to that presented above, based on the conditions for the absolute stability of onions for two-cell types and medium and three-cell types and medium.

4. LOCALLY STABLE PATTERNS

The patterns so far discussed have been absolutely stable, that is, they have the maximum *E* value for any possible pattern formed by cells having the assigned λ values. We call a pattern locally stable if its *E* value, even though not maximal, is greater than the *E* value of any neighboring pattern. A neighboring pattern is defined as one that can be reached from the given pattern according to some cell motility rule. Because the real motility rules of cells are not definitely known, some apparently plausible rules will be assumed.

Given a set of motility rules, the number of possible locally stable patterns is large, and thus a general study of locally stable patterns is difficult and will not be attempted here. Attention will be concentrated on the conditions for local stability of a few structures of biological interest. leaving aside the question of how such locally stable patterns were formed in the first place. It is thus obvious that, whereas the study of absolutely stable patterns has a fairly direct bearing, in spite of the simplifications of the model, on the possible formation of real biological structures through the operation of cell motility and cell adhesion, a similar study of locally stable patterns is less likely to represent biological reality until the actual rules of cell motility become better known. Nevertheless, model studies of locally stable patterns are worthwhile because they throw some light on the nature of the problem. From these remarks it is clear that the local stability of a pattern depends strongly on the motility rules because such rules determine what is and what is not a neighboring configuration.

It is realistic to assume that a cell is restricted in its movements; here we will assume that a cell can only move to a neighboring square of the tessellation. Let us say that two patterns, P_1 and P_2, differ by an elementary permutation if they are identical except that the contents of two laterally or diagonally adjacent cells are interchanged. Two elementary permutations are called compatible if they involve no overlapping of squares of the tessellation. We shall say that two patterns are neighboring if they differ by a number of compatible elementary permutations. Thus any individual pattern P_0 determines a set of other patterns that are its neighbors. It is useful to think of time as moving along a discrete set of instants so that a cell is either in one square of a tessellation or in another. Within this unit of time, the value *E* of a pattern does not change: in the next unit of time, it may be the same or different.

Unfortunately the choice of rules governing cell movement is very wide. In attempting to put some restrictions on the choice, one may suggest that movement will be such that the E value of the subsequent pattern will always increase or at least remain the same. In some sense this requirement is implied by the experimental results: configurations of maximum E do not change into submaximal configurations with time. So far as elementary permutations are concerned, this restriction may be interpreted to mean that the requirement is global, that is, the E of the overall pattern must increase even if some moves are locally energetically unfavorable. Because this conclusion implies that individual cells have some way of detecting and responding to information as to the global pattern of which they are a part, it does not seem very plausible. The alternative is to assume that cells will only move from an environment of locally lower E to an environment of locally higher E. Somewhat less restrictive would be the assumption that moves to equal E are permitted, or even moves to lower E with reduced probability.

We discuss the nature of the real motility rules later. For the purpose of studying local stability we shall assume merely that a cell can undergo a single flip between adjacent diagonal or nondiagonal squares of the tessellation, and individual flips will occur only when the individual flip causes an increase in E of the total pattern.

The structures whose local stability we investigate by using the above motility rules are the histologically significant ones, that is single layers, either as sheets (epithelia), or tubules and vesicles. In two dimensions, they may be represented by layers and rings (or rectangles) of cells. Owing to the nature of our grid, tubules and vesicles are represented by rectangular structures; this method introduces some artifacts because of the presence of corners that do not appear in the structures they represent. A more serious criticism is that, because the figures we study are cross sections of three-dimensional figures, they are ambiguous. A ring may represent the cross section of a sphere, a tubule, and so on. These considerations may be regarded as making the similarity between our models and actual three-dimensional figures quite remote. This interpretation is indeed true but we present our results because they are an entering wedge to more realistic representations. Let us discuss separately the local stability of patterns involving two and three types of cells.

A. Patterns Involving Two Types of Cells

Let us call the patterns whose local stability we wish to study P, and denote the neighboring patterns by P'. From the remarks made earlier in connection with the absolute stability of patterns with two types of cells, it is clear that for the study of local stability in a two-cell system, it is enough to know the difference between the number of any one of three types of edges in P and P'.

Let us choose *bb* as this type. Allow us also to assume $N_w \gg N_b$. Set

$$N_{bb}(P) - N_{bb}(P') = M_1. \tag{46}$$

Therefore, from equation (9a),

$$E(P) - E(P') = M_1\mu_1. \tag{47}$$

Because for any P of biological interest we can always find a P' for which $M_1 > 0$ (namely, by interchanging a b cell with a w cell), for P to be locally stable we must have

$$\mu_1 = \lambda_{bb} + \lambda_{ww} - 2\lambda_{bw} > 0. \tag{48}$$

This is the first necessary condition for the local stability of P. The other necessary condition is that there be no P' for which $M_1 \leqslant 0$. Together, these two conditions are necessary and sufficient for the local stability of P. If there exists a P' for which $M_1 = 0$ and no P' for which $M_1 < 0$, P is a metastable pattern. We now discuss the local stability of rings and sheets of cells.

Rings of Cells

For convenience we divide these rings into three categories and discuss them separately.

Single-layered ring. The simplest pattern in this category has eight b cells, $N_b = 8$, with one w cell trapped inside [fig. 4(a)]. This pattern is unstable because, for the pattern P' obtained from this pattern P by interchanging the trapped w with any b, $M_1 < 0$.

The next simplest pattern is the one with more w cells inside but arranged in the form of a layer one cell thick [fig. 4(b)]. This pattern is metastable if we

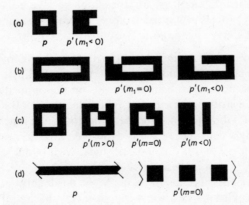

Fig. 4. Single-layered rings [(a), (b), (c)] and infinite sheet of cells (d). (a) $N_b = 8$; (b) Long ring; (c) $N_b = 12$. For each case P' is obtained from P by a single non-diagonal flip except for (c). For (c) $P'(M = 0)$ is obtained by single diagonal flip, $P'(M < 0)$ is obtained by two diagonal flips.

allow only nondiagonal flips because, for P' obtained from P by interchanging an end internal w cell with any neighboring b cell, $M_1 = 0$. If we also allow diagonal flips, it becomes unstable because, for the pattern P' obtained from P by interchanging an end internal w cell with a corner B cell, $M_1 < 0$. This illustrates the sensitivity of local stability to the motility rules.

In the next pattern that we consider, the w-cell layer is two cells thick. The simplest such pattern is $N_b = 12$, with four w cells trapped inside [fig. 4(c)]. This pattern is stable if we allow only a single nondiagonal flip because there is no P' for which $M_1 < 0$. But, if we allow a diagonal flip, this pattern becomes metastable because, for example, for the pattern P' obtained by interchanging a corner b cell with a diagonally opposite w cell, $M_1 = 0$. On the other hand, if we allow simultaneous compatible flips, the same pattern becomes unstable (for P' obtained by interchanging the two w cells by their diagonally opposite b cells, $M_1 < 0$). Even if $N_w > 4$, as long as w cells are arranged in layers two cells thick, the results for the case $N_w = 4$ hold good.

We now consider single b-layered patterns with inner w cells arranged in at least three cell layers. These patterns are stable if only a single nondiagonal flip is allowed. They are metastable if a diagonal flip is also allowed because, if we take one corner w cell inside P_1 and interchange it with the diagonal b corner cell, the resulting pattern P' has $M_1 = 0$. Similar results hold for larger single-layered patterns.

From the above discussion it is clear that to obtain a pattern P' so that $M_1 < 0$, we must be able to interchange b and w cells in such a fashion that the b cells on the opposite edges of P come in contact with one another. This is not possible with our motility rules with only a single flip allowed for single b-layered patterns (with the exception of those in which inner w cells are arranged in the form of layers one cell thick). However, if we allow more than one flip until E changes, all the single b-layered patterns become unstable because, for P' obtained from P by interchanging w cells on the edges of the inside layer by diagonally opposite b cells, $M_1 < 0$.

Two-layered rings. The simplest pattern in this category will have one w cell inside two layers of b cells. This pattern is not locally unstable with our neighbor rules because, even bringing b cells on the opposite sides of a w cell into contact, there is no increase in N_{bb}. This one-w-cell pattern is a metastable pattern because P' obtained by interchanging any of the inner w cells with any of its neighboring b cells has $M_1 = 0$. However, a pattern with inner w cells arranged in the form of a layer one cell thick, with two b layers, is unstable if we allow more than one consecutive flip.

The next pattern in this category with $N_w = 4$ is locally stable because any interchange of w with b will always decrease N_{bb}. In fact, if the inner w cell layer is at least two cells thick, the two b-layered rings will always be locally stable. We may note that the local stability of two b-layered rings does not change whether we do or do not allow diagonal flips.

Three- or more-layered rings. We leave as a trivial exercise for the readers to show that whatever we have said for two-layered rings is completely valid for this case.

Sheets of cells

A sheet is represented as a chain of b cells in contact with w cells on both sides [fig. 4(d)]. If we allow only one flip, this chain P is stable because, for any P' obtained by any interchange, $M_1 > 0$. However, if we allow simultaneous compatible flips, the chain will be unstable if it is finite and will tend to break up into the pattern

$$
\begin{array}{c}
..b-b.....b-b.....b-b.. \\
\mid \ \ \mid \qquad \mid \ \ \mid \qquad \mid \ \ \mid \\
b-b \ \ \ \ \ b-b \ \ \ \ \ b-b
\end{array}
$$

If the chain is infinitely long, rearrangement into the broken form is a metastable transition ($M_1 = 0$) [fig. 4(d)]. This result is in accordance with the observed fact that, if an epithelium layer is stretched beyond a certain point, the layer disappears. If the sheet is two cells thick, any flip will decrease bb contacts and hence is locally stable.

B. Patterns Involving Three Types of Cells

Analysis of a system of three types of cells requires three μs and 17 different forms of E. This makes the general analysis of local stability rather complicated. Finding the set of values of μ_i that will make a particular pattern locally stable for a given set of motility rules involves calculating $E(P) - E(P')$ for all P' in 17 different forms and then finding the values of μs for which $E(P) - E(P') > 0$ for all P'. We will not do this analysis in this paper; instead, we will find the set of μs for which the locally stable patterns for two types of cells will remain stable and for which the unstable or metastable ones either remain so or become more stable. We discuss the rings and sheets of cells.

Rings of cells

Let us denote $N_{bb}(P) - N_{bb}(P')$, $N_{ww}(P) - N_{ww}(P')$, and $N_{ee}(P) - N_{ee}(P')$ by M_1, M_2, and M_3, respectively. For a ring pattern for any flip, $M_3 \geqslant 0$. For a capillary ring pattern, that is a pattern in which w cells are arranged in the form of a layer one cell thick, any flip between a pair of neighboring b and w cells that are not diagonal to each other will give P' with $M_1, M_2 > 0$, and $M_3 = 0$, and the flip between diagonal cells will give P' with $M_1 = M_3 = 0$, $M_2 > 0$. For noncapillary rings, independent of the thickness of b layers, for all P', $M_2 > 0$ and $M_1 \geqslant 0$. From form (8) of E (table 1),

$$
E(P) - E(P') = M_1(\mu_1 + \mu_2 - \mu_3) + M_2(\mu_1 + \mu_3 - \mu_2)
$$
$$
+ M_3(\mu_2 + \mu_3 - \mu_1). \tag{49}
$$

Because for ring patterns our motility rules allow only $M_1 \geqslant 0$, $M_2 > 0$, $M_3 \geqslant 0$ if

$$\mu_1 + \mu_2 > \mu_3 \qquad \text{i.e.} \qquad \mu_{bb} + \mu_{we} > \mu_{be} + \mu_{bw} \qquad (50a)$$

$$\mu_1 + \mu_3 \geqslant \mu_2 \qquad \text{i.e.} \qquad \mu_{ww} + \mu_{be} \geqslant \mu_{bw} + \mu_{we} \qquad (50b)$$

$$\mu_2 + \mu_3 \geqslant \mu_1 \qquad \text{i.e.} \qquad \mu_{bw} + \mu_{ee} \geqslant \mu_{we} + \mu_{be}, \qquad (50c)$$

all these ring patterns that were stable for the two-cell type will remain stable and those that were unstable or metastable will either remain so or become more stable.

Thus, by bringing another type of cells for suitable μs, we have made single b-layered rings also stable. This development is not very surprising because by introducing another type of cell we have artificially introduced anisotropy in the cell surface, that is, two edges of b cells are in contact with b cells, one with a w cell and one with an e cell. We had another example of stability due to anisotropy in the system with two types of cells where the double b-layered ring became stable. As shown elsewhere (Goel and Leith, 1970), the anisotropy can make epithelial, vesicular, and tubular structures absolutely stable.

We conclude the discussion on the local stability of patterns involving three types of cells by showing that we can make single b-layered rings more stable than complete and partial multilayered b rings for the same number of w cells by choosing appropriate values of λs. To find these values we note that the maximum number of b cells that can be accommodated around w cells in a single-layer ring is

$$N_b = 2N_w + 6. \qquad (51)$$

Therefore, for the purpose of the present discussion, the relevant number of b cells is $N_b \leqslant 2N_w + 6$. For single or multilayer structure, $N_{we} = 0$. Let us choose form (7) of the E function, that is,

$$E = N_{bb}(\lambda_{bb} - 2\lambda_{be} + \lambda_{ee}) + N_{ww}(\lambda_{ww} - 2\lambda_{bw} + 2\lambda_{be} - \lambda_{ee}). \qquad (52)$$

Because for a multilayered b ring, both N_{bb} and N_{ww} are greater than those for a single b-layered ring, the sufficient conditions for a single b-layered ring to be most stable among b-ring structures, with w cells inside, are

$$\lambda_{bb} < (2\lambda_{be} - \lambda_{ee})$$

$$\lambda_{ww} + (2\lambda_{be} - \lambda_{ee}) < 2\lambda_{bw}.$$

It may be noted that these conditions imply

$$\lambda_{bb} + \lambda_{ww} < 2\lambda_{bw}.$$

We may add that our one-layer pattern is metastable, that is, there is more than one b-layered pattern (although topologically identical) that has the same energy. One such pattern is (for $N_w = 4$) as shown:

bbbbb

bwwwb

bbbwb

bbb

Sheets of cells

Given any of a number of cell motility rules, we find that single layers of cells with the same environment on both sides are locally unstable. However, two or more layers are locally stable. It is therefore of interest that in general when sheets of cells occur, they do not have the same environment on both sides of the sheet. This is the case for all epithelia. When there is a sheet of cells with the same environment on both sides, the sheet is double.

It may be noted that this generalization and in fact much of the discussion on cell sorting has little relevance to plant histology. In plants the cells are held rigidly with respect to one another by cell walls, and much morphogenesis is a matter of the pattern of cell division.

Although these cases of local stability are primarily presented only for illustration, it is worthwhile to compare the conclusions with histological observations. We consider two cases, the occurrence of single and double sheets of cells and the reaggregations of embryonic lung tissue.

C. Reaggregation of Embryonic Lung Tissue

Most experimental work on vertebrate reaggregation has dealt with the formation of "onions." Studies on the appearance of more complex patterns in reaggregates (Moscona, 1965) have not been detailed. However, Grover (1961, 1962, 1963) has described in considerable detail the reaggregation of dissociated embryonic chick lung tissue, which is of interest to consider from the point of view of our model. Dissociated embryonic lung contains mainly two cell types, epithelial and mesenchymal. [We will not discuss here the vascular tissue although it may affect reaggregation pattern (Grover, 1962).] The course of reaggregation in a hanging drop depends on the relative proportions of the two types of cells.

When the proportions are optimal, reaggregation produces masses of mesenchyme coated with a single layer of epithelial cells. Within the mesenchyme are islands of epithelial cells that eventually form vesicles and tubules lined with epithelium.* This formation produces a rough outline of the

---

* Tubular structures are rare in reaggregates of 7-day lung but develop well from cells of lung 11 to 12 days old.

embryonic lung structure, shown schematically in figure 5. In terms of our model, this configuration can be represented as in figure 6, and reaggregation as in figure 7.

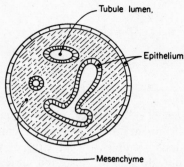

Fig. 5. Schematic representation of cell configuration in lung aggregates as described by Grover (1961, 1963).

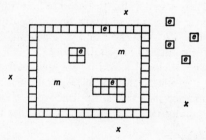

Fig. 6. Model representation of lung reaggregate. It appears that tubule lumens appear as a secondary cavitation of epithelial cell masses, and here we deal only with the initial aggregation. Excess epithelial cells are observed to remain isolated in the medium (upper right); m = mesenchymal cells, e = epithelial cells, x = medium.

To analyze this behavior in terms of our model, we can consider as a standard Grover's major pattern, a mesenchymal mass covered with epithelium and containing an epithelium-lined tubular system. We then consider the energy changes associated with various transformations away from this pattern.

It is useful to express the energy E of a pattern containing three cell types m-mesenchymal, e-epithelial, x-medium as

$$2E = -\mu_1 N_{me} - \mu_2 N_{ex} - \mu_3 N_{xm}, \tag{53}$$

where

$$\mu_1 = \lambda_{mm} + \lambda_{ee} - 2\lambda_{me} \tag{54a}$$

$$\mu_2 = \lambda_{ee} + \lambda_{xx} - 2\lambda_{ex} \tag{54b}$$

$$\mu_3 = \lambda_{xx} + \lambda_{mm} - 2\lambda_{xm}. \tag{54c}$$

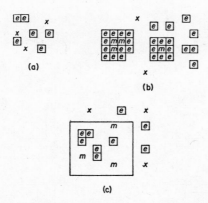

Fig. 7. Reaggregation of lung. (a) Pure epithelial cells. (b) Epithelial cells with small amounts of mesenchyme. (c) Excess mesenchyme with a few epithelial cells. *m, e* and *x* have same meanings as in Fig. 6.

Let us first consider the stability of the major pattern that Grover observed, namely, a mesenchymal mass coated with epithelial cells and additional epithelial cells in suspension and see if we can choose a set of μs such that this pattern (P) is more stable than other patterns (P'). Let

$$2\Delta E = 2[E(P) - E(P')] = \mu_1 n_1 + \mu_2 n_2 + \mu_3 n_3, \tag{55}$$

where

$$n_1 \equiv N_{me}(P') - N_{me}(P) \tag{56a}$$

$$n_2 \equiv N_{ex}(P') - N_{ex}(P) \tag{56b}$$

$$n_3 \equiv N_{xm}(P') - N_{xm}(P). \tag{56c}$$

We require $\Delta E > 0$ for all P'. For our pattern, $N_{xm}(P) = 0$. Let us first consider patterns for P' for which $N_{xm}(P') = 0$, so that $n_3 = 0$. Some of the typical P' are as shown in figure 8(a)–(f). Pattern (a) is formed when additional e cells are recruited from the medium and $m - e$ interaction is increased by the formation of a checkerboard between m and e. Pattern (b) is formed when sufficient e cells will be recruited from the medium to completely surround each m cell with clusters each of one m-cell and four e-cells left suspended in the medium. For both P and $P', n_1, n_2 > 0$, but n_2 is not too much greater than zero. Pattern (c) is also formed by a similar process. Patterns (d)–(f) involve changes of one or two cell positions. Pattern (d) has a preliminary checkerboard configuration; (e) involves a suspended cell "precipitating" onto the structure to make the multilayered epithelium, and (f) involves an increase of epithelial area through recruitment of cells from suspension with consequent distortion of the mass. For the patterns (c)–(f), $n_1 > 0$ and $n_2 < 0$. If we choose $\mu_1 > 0$ and $\mu_2 \simeq 0$, then $\Delta E > 0$ for all P' shown in figure 8(a)–(f).

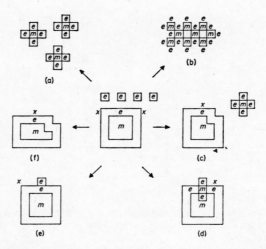

Fig. 8. Typical possible neighboring-type patterns [(a) to (f)] from basic epithelial-covered mesenchymal mass (center).

We now consider those P' for which $N_{xm} \neq 0$. For such P' [fig. 9(a)] at least a single e cell should leave the continuous epithelium, exposing one M cell to the medium. For pattern (a) of figure 9, $n_1 = -1$, $n_2 = 5$, $n_3 = 1$. We have to choose μ_is so that $-\mu_1 + 5\mu_2 + \mu_3 > 0$. This choice has to be consistent with the condition derived above, namely, $\mu_1 > 0$, $\mu_2 \simeq 0$. Such a choice could be $\mu_1 = 5$, $\mu_2 = 1$, $\mu_3 = 1$. It may be noted that this choice will energetically allow the dissolution of a discontinuous epithelium, as apparently occurs. These dissolutions are shown in figure 9(b) and (c), the trans-

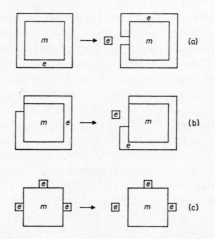

Fig. 9. Possible transformations of basic pattern allowing $\Delta N_{mx} > 0$.

formation leading to (b) corresponds to $n_1 = 1, n_2 = -3, n_3 = -1$, and the one leading to (c) corresponds to $n_1 = 1, n_2 = -1, n_3 = -1$.

The choice of μs in this fashion explains almost all Grover's observed reaggregation patterns as patterns of local stability. It stabilizes the general pattern of an epithelium coating a mesenchymal mass and provides for excess epithelial cells remaining in the medium in a loosely associated fashion. These epithelial cells will not intercalate into the epithelium to distort the masses from a near-spherical shape. However, these cells would add on to existing epithelia with slight increase in energy; it is necessary to assume that, for some reason, suspended cells are unable to add to the mass. Thus the assumptions that cell properties do not change with time, and that the cells are isotropic so far as their adhesive properties are concerned, seem to suffice to explain most, but not all, of the phenomena of reaggregating lung.

In this and in other cases it is likely that sorting out is the major process in early pattern formation, but it is apparent that at some point simple reaggregation processes end and tissues begin to differentiate by virtue of changes in the cell properties. Presumptive tubule cells secrete fluid, for example, and in this way change in configuration from a clump to a shell. This process involves cells secreting in a polarized fashion and it also apparently involves the formation of tight junctions (Wood, 1965; Wartiovaara, 1966). Clearly the events beyond the initial stages must be considered in terms of modified cell properties.

D. The Concept of an Epithelium

All the histological structures whose local stability has been considered are essentially single sheets of cells, sometimes closed into tubules or vesicles. They are thus examples of epithelial structures. We wish to stress at this point that we define an epithelium as something more than merely a single layer of cells spread over an underlying mass of another type of cells. By an epithelium we mean a single cell layer with an intrinsic tendency to remain so when more cells are added to it. The additional cells will not add to the original single layer but rather will intercalate into the original layer. To accommodate them, the epithelial cells may be distorted, or the underlying cell mass may be distorted to increase its surface-to-volume ratio. The difference between a true epithelium and a fortuitous single cell layer, resulting from a restriction on the number of cells present, is illustrated in figure 10.

Now as we have shown, an epithelium is not an absolutely stable structure if the cells are isotropic, although it may be a locally stable configuration. Some further investigations by Goel and Leith (1970) have shown, however, that if cells are nonisotropic, having μs that vary with direction, an epithelium is one of the absolutely stable structures. In fact, it is very likely that epithelial cells are nonisotropic as is suggested, for example, by the polarized formations of basal lamellae or adhesive junctional complexes. This situation would

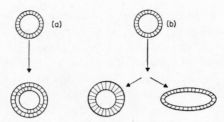

Fig. 10. (a) is fortuitous single-cell layer structure because addition of cells of the type forming the single layer results in the formation of additional layers; (b) is a true epithelium because added cells results either in the formation of columnar epithelium or distortion of underlying cell mass; in either case, the outer surface remains a single layer of cells.

probably make it possible to greatly improve the concordance between the model and such observations as the reaggregation of lung. It is nevertheless of interest that even local stability goes some way to explain the persistence of certain histological structures.

5. The Path Problem and Cell Mobility

Steinberg's hypothesis provides a method by which cells can produce a maximum E configuration from other configurations, including random ones. Nevertheless, *a priori*, it is by no means obvious that such patterns can be reached from any configurations. Indeed, the experimental results show that under certain conditions patterns are reached that are locally, but not absolutely, stable. One therefore has to investigate the requirements that have to be met when pattern P_2 has to be reached from pattern P_1. This is the general path problem. Here we consider a restricted part of this problem: Starting with a random distribution of cells, Which motility rules make it possible to reach a configuration of concentric spheres? Another part of the problem is as follows: Given a set of motility rules, in general, cells can follow more than one path when pattern P_2 is to be reached from pattern P_1. Some of these paths are shorter than others, that is, some require a smaller number of cell movements than others. Further, in general, in different paths the value of the E function will change differently with every cell movement. The problem is, Which of the paths are the cells going to follow in real cell sorting? We will not discuss this part of the problem. Parenthetically, it may be remarked that finding a histological pattern of maximal E does not logically imply that it has been produced by a sorting out of cells.

One can consider an alternative generating procedure that consists of adding or subtracting cells from a maximal pattern but of a composition that is different from that of the given pattern. Thus, suppose $C(N_b, N_w, N_e)$ consists of the class of patterns each of which has N_b cells of type b, N_w cells of type w, and N_e cells of type e (the medium cells). Given pattern P maximal in $C(N_b, N_w, N_e)$, it is possible that P can be obtained from a maximal pattern Q in $C(N_b - M, N_w + M, N_e)$ by simply replacing Mw cells in Q by Mb cells. Such a replacement can be viewed as a differentiation of w cells into b cells, but the question where this differentiation occurs in Q is dictated by the requirement that the property of maximality shall be retained. The particular w cells that can be converted into b cells while preserving maximality can be characterized by a bond energy property if the bond energy coefficients λ_{bb}, λ_{bw}, and λ_{ww} satisfy certain reasonable relations, as discussed in section 3. Such a property consists of the w cell having maximum bond energy relative to all other w cells in the pattern Q.

Here, however, we wish to consider the possibility of reaching a stable pattern through cell movement alone. In tems of elementary permutation, as previously described, cell movements must be such that at least statistically the value of E of a pattern increases or at least remains the same. This requirement is implied by the experimental findings: configurations of maximum E do not change with time.

Unfortunately, given this constraint, the choice of rules governing cell movement remains very wide. Two general classes of rules are possible. The movement of a cell may be governed by the requirement that the E value of the entire pattern increases or at least remains the same. This global requirement implies that individual cells have some way of sensing what the global E value is. Physically this explanation does not seem very plausible. It seems more plausible to assume that a cell is only aware of local changes in E values. It will then move from a lower to a higher E configuration. Somewhat less restrictive would be the assumption that moves to configurations of equal E be permitted, or even moves to lower E; either of them represents the only possible moves, or permitted moves but with lower probability.

Although such motility rules appear plausible *a priori*, we have found that they are not adequate to explain the experimental findings. We ask here whether concentric spheres, a configuration of maximum stability, can usually be reached from a random distribution of cells because this is the experimental finding. The postulated motility rules will be clearly inadequate if they do not lead to such a result.

We have investigated the adequacy of some motility rules through computer simulation by assuming that cells will move only to higher E configurations, to configurations of equal E, or to configurations of lower E if no other moves are possible.

Using a random number generator, b cells were placed on a 20×20 array in a FORTRAN program at random with probability f_b. f_b is thus the density of b cells in the tessellation. The rest of the cells were labeled w. A strip of w cells, two wide around the array, was used to confine the b cells [fig. 11(a)]. All the b cells were scanned one at a time during each computing cycle (or time step), but in random order so as to avoid systematic effects. For

Fig. 11. (a) Starting array of b (black) and w (white) cells with $f_b = 0.58$. (b), (c) and (d) are patterns generated by computer simulation for $\Delta E_{min} = 1$, 0 and -1, respectively, after 100 scans of b cells.

each b cell, a list of its eight nearest and next-nearest neighboring w cells was constructed. The increase in nearest neighbor b–b pairs, ΔE, that would occur was then calculated for each possible switch. To calculate ΔE, λ_{bb} was assigned the value of 1, λ_{ww} and λ_{bw} that of 0. If $\Delta E < \Delta E_{min} = 1$, the switch was rejected. If all neighboring w cells were rejected, the b cell was not moved. If any w cells remained, for which $\Delta E \geqslant 1$, then the one for which ΔE was maximal was chosen. If two or more w cells had the maximal ΔE, one of them was chosen at random. The switch was made and the program went to the next b cell. The computing cycles were stopped after each b cell had 100 opportunities to move or after a scan produced no switches. (With $\Delta E_{min} = 1$, generally each b cell switched no more than twice). The array was then printed, along with counts of b–b nearest neighbor pairs and b clusters. (A cluster or clump was defined as a connected set of b cells in which the connections are between nearest neighbors.) The pattern resulting from the cell sorting is as shown in figure 11(b), and the average number of cells per clump is plotted vs. density of b cells in figure 12, both for the starting random arrays and the patterns resulting from cell sorting. With $\Delta E_{min} = 1$, single clumps are obtained when $f_b \gtrsim 0.7$. In figure 13 we have plotted N_{bb}/N_b

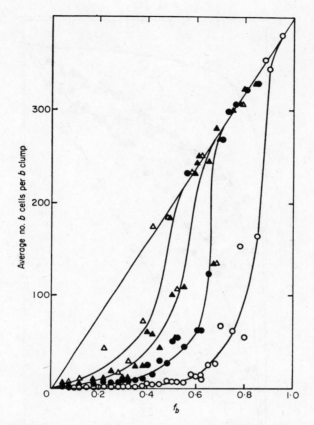

Fig. 12. Average number of b cells per b clump in a 20×20 array vs. the fraction of b cells f_b. (○) initial random array, (●) $\Delta E_{\min} = 1$, (▲) $\Delta E_{\min} = 0$, (△) $\Delta E_{\min} = -1$. On the straight line, all the cells would be in a single clump. Some points are above this line because of fluctuations above the expectation for N_b.

vs. f_b for this pattern resulting from the self-sorting of cells and also for the most stable configuration (top centre). From this figure we see that these clumps are not in the most stable configurations. Also, they are not compact and often include trapped w cells [fig. 11(b)].

We thought the situation would improve if we did not insist that N_{bb} increase for every switch. Thus we set $\Delta E_{\min} = 0$. The resulting pattern is shown in figure 11(c). Single clumps were now formed when $f_b \gtrsim 0.6$ (fig. 12), but again, although N_{bb}/N_b improved over the previous case (fig. 13), the clumps were still far from the most stable configuration. We then set $\Delta E_{\min} = -1$, which allows a cell to make a slightly unfavorable move if that is all that is available. The resulting pattern is shown in figure 11(d). Single clumps were obtained when $f_b \gtrsim 0.5$ (fig. 12) but N_{bb}/N_b did not

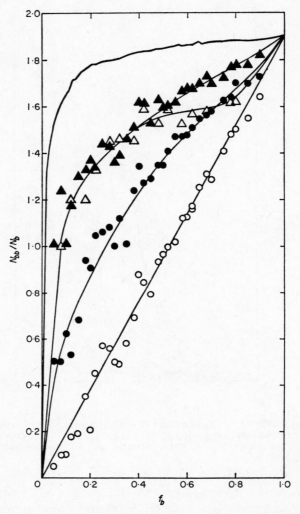

Fig. 13. N_{bb}/N_b vs. f_b for the same computer runs as for Fig. 12. The straight line is the expectation for the initial random arrays. The fluctuation of the points around this line is an indication of the accuracy of the Monte Carlo calculations. The curve at the top is the maximum possible value of N_{bb}/N_b. The curve is plotted by using the relation $N_{bb} = 2[N_b - N_b^{\frac{1}{2}}]$ where the brackets signify truncation to an integer.

improve at all (fig. 13). $\Delta E_{min} = -2$ and -3 gave worse results, probably because of too much evaporation of b cells from b clusters. Perfect square clusters, used as starting conditions, acquired rough edges and even broke up when $\Delta E_{min} \leqslant -1$. For these rules a square is not locally stable.

At this point we turned to three dimensions to see if such a context made it possible to reach the absolutely stable onion structure. We wrote another FORTRAN program, simulating a $10 \times 10 \times 10$ array of cells and accounting for all 26 first, second, and third nearest neighbors. The results were qualitatively similar to the two-dimensional cases. Although the transitions to single clumps occurred at much lower values of f_b (fig. 14), there was still a

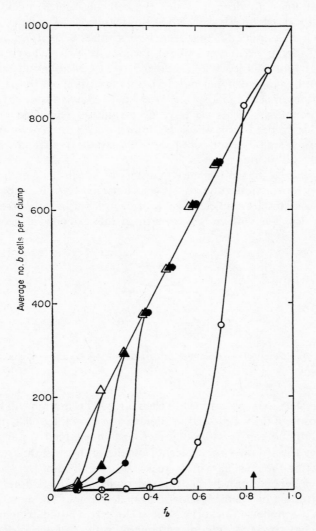

Fig. 14. Average number of b cells per b clump, in a $10 \times 10 \times 10$ array, vs. f_b. The arrow at $f_b = 0.831$ indicates the critical percolation probability for a cubic lattice (Hammersly & Handscomb, 1965). The key is the same as in Fig. 12.

considerable excess of *b–w* pairs. Cross sections of the three-dimensional clusters looked much like the two-dimensional clumps (fig. 11). On both lattices, $\Delta E_{min} = 1$, presumably closest to describing the behavior of real cells, gave the worst sorting out.

Clearly our motility rules, which at first seemed quite reasonable, are unable to give the onion configuration, a clump of one type of cells enclosed by another, for a random mixture; the cell aggregates quickly become trapped in local maxima. It is true that this will occur to some extent in real systems, but in our model this tendency is much too pronounced: considerable trapping occurs for all proportions of two cell types. Clearly, Steinberg's hypothesis needs some reformulation. Cell adhesion and unspecified cell motility are not sufficient to explain the sorting out that is observed: the cell motility must be of a special kind that we have not discovered.

In the preliminary trials we have by no means exhausted all possible motility rules, so pessimism would be unwarranted at this point. We intend to continue the study. As an example of a situation we have not so far considered is the following.

It is possible to regard the initial clusters of *b* cells as a liquid network undergoing contractions due to surface tension. Consider, for instance, two clusters of *b* cells connected by a bridge of *b* cells, all surrounded by *w* cells (fig. 15). The surface tension around this extended drop will cause

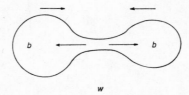

Fig. 15. Schematic diagram of two clumps of *b* cells connected by a bridge of *b* cells, all embedded in *w* cells.

the two *b* clusters to be pulled toward each other. This motion will be resisted by the viscosity of the *w* cells and the deformability of the *b* cells themselves. On the other hand, this very same surface tension will tend to pull the *b* cells in the bridge toward the two clusters. Thus the bridge will thin out and the force attracting the clusters will decrease. Whether or not the two clusters will fuse then depends on whether they are pulled together before the bridge snaps. The timing of interdependent events determines the pattern formed. Dynamically, a model of this sort would be closer to a liquid drop model than the ones we have been considering and might lead to a closer approximation of the experimentally observed results. Observations by Trinkaus (1969) on retina–heart cell sorting support this viewpoint in many details.

6. Self-Assembly and Adaptive Systems

Our model, as stated before, can be elaborated in a variety of ways to correspond more closely to reality and, hopefully, to make it useful as a predictive tool. Its interest is enhanced if it is viewed as part of a larger problem, that of self-assembly or self-organization. Self-assembly has already received considerable attention in two fields of biology, that of the folding of polypeptide chains and the assembly of virus from its component proteins and nucleic acid. A nonbiological system that is closely related to our model is the behavior of mixtures of immiscible fluids (Torza and Mason, 1969). Such a system is characterized by the coalescence of various liquid phases to produce a uniquely determined final configuration. All these examples of self-organization share certain features with our model of cell self-assembly. First, each system can exist in a large number of alternate configurations or states. Second, the systems are motile in that they are free to explore alternate configurations. Third, there is a measure of system performance available according to which the system moves to improve its performance with respect to this measure. In our case this measure is given by an analog of free energy associated with each configuration. In this sense all the systems behave adaptively, which is the reason there is such an overlap between the literature on self-organization and that on adaptive systems.

This work was supported by the National Aeronautics and Space Administration and by the American Institute of Biological Sciences at the Colloquium on Theoretical Biology held in Fort Collins, Colorado in July and August 1969.

REFERENCES

Goel, N. S. and Leith, A. G. (1970) *J. Theor. Biol.* 28: 469.
Grover, J. W. (1961) *Develop. Biol.* 3: 555.
Grover, J. W. (1962) *Exp. Cell. Res.* 26: 324.
Grover, J. W. (1963) *Nat. Cancer Inst. Monograph* 11: 35.
Hammersly, J. M. and Handscomb, D. C. (1965) *Monte Carlo Methods.* London: Methuen.
Moscona, A. A. (1965) In *Cells and Tissues in Culture*, vol. 1, ed. E. N. Willmer. New York: Academic Press.
Steinberg, M. S. (1963) *Science, N.Y.* 141: 401.
Steinberg, M. S. (1964) In *Cellular Membranes in Development*. ed. M. Locke. New York: Academic Press.
Torza, S. and Mason, S. G. (1969) *Science N.Y.* 163: 813.
Townes, P. L. and Holtfreter, J. (1955) *J. Exp. Zool.* 122: 53.
Trinkaus, J. P. (1969) *Cells into Organs.* New Jersey: Prentice-Hall.
Wartiovaara, J. (1966) *Ann. Med. Exp. Biol. Fenniae Helsinki* 44: 469.
Wilson, E. B. (1907) *J. Exp. Zool.* 11: 245.
Wood, R. L. (1965) *Anat. Record* 151: 507.

APPENDIX

Boundary Effects in Relation to the Stability of Patterns
Involving Two Types of Cells

In this appendix we will reformulate the problem of study of patterns involving two types of cells allowing any number of cells of any type to go to the boundary. Let us treat the vacuum around the mass of cells as another type of cell (e type).

We demand our pattern to be continuous so that N_{ee} remains constant. We can then use the formalism developed in section 3 for three cell types with the restrictions N_{ee} constant, $\lambda_{be} = \lambda_{we} = \lambda_{ee} = 0$. Let us first discuss absolute stability.

If we use form (1) (table 1) of E, that is,

$$E = N_{bb}\lambda_{bb} + N_{bw}\lambda_{bw} + N_{ww}\lambda_{ww} + \text{constant}, \tag{A1}$$

then for $0 \lesssim \lambda_{bw} \ll \lambda_{ww} \lesssim \lambda_{bb}$, the absolutely stable pattern is the one in which b and w are completely segregated and touching only at one point. On the other hand, if we use form (6), that is,

$$E = N_{bb}(\lambda_{bb} - 2\lambda_{bw} + \lambda_{ww}) + N_{be}(\lambda_{ww} - \lambda_{bw}) + \text{constant}, \tag{A2}$$

then for

$$\lambda_{bb} + \lambda_{ww} - 2\lambda_{bw} > 0 \qquad \text{i.e. } \mu_1 > 0 \tag{A3}$$

and

$$\lambda_{bw} > \lambda_{ww}, \tag{A4}$$

the onion structure with b cell inside will be absolutely stable. Note that if $N_{be} = 0$, that is, b cells are not allowed to go to the boundary of pattern with vacuum, equation (A1) reduces to equation (9a).

Let us now use form (8), that is,

$$E = N_{bb}(\lambda_{bb} - \lambda_{bw}) + N_{ww}(\lambda_{ww} - \lambda_{bw}) + \text{constant}.$$

Then for

$$\lambda_{bb} < \lambda_{bw}, \qquad \lambda_{ww} < \lambda_{bw},$$

the pattern with minimum N_{bb} and N_{ww}, that is, checkerboard patterns, will be absolutely stable.

It may be noted that the local stability results are not affected, because we have excess of w cells and, according to our neighbor rules, b cells are not allowed to go to the boundary.

8. Self-Sorting of Anisotropic Cells

Narendra S. Goel and Ardean G. Leith*

In the preceding paper Goel et al., by using a simple model that assumes (a) differential affinities between cells of different types and (b) local motility of cells, showed that, for a self-sorting system of isotropic cells, none of the absolutely stable configurations is a two-dimensional analog of biologically interesting structure, for example, tubular, epithelial, or vesicular. Using the same model, it is shown here that some of the absolutely stable configurations for the two-dimensional anisotropic cells are the analogs of these biologically interesting structures.

1. INTRODUCTION

In the preceding paper (Goel et al., 1970), we examined some of the consequences of a hypothesis proposed by Steinberg (1963, 1964). According to this hypothesis, if cells possessed what might be called differential affinities between cells of different types (including different affinities for the medium) and if the cells were at least locally motile, they would automatically tend to sort themselves out in such a manner as to produce structures of maximal stability. We will state shortly exactly what is meant by maximal stability, but roughly it means maximum number of contacts between cells with the highest degree of mutual adhesiveness. One consequence of this hypothesis that we examined in the preceding paper is the types of maximally stable configurations that isotropic nondeformable cells will make.

To determine these configurations, we modeled the system of isotropic nondeformable cells as a two-dimensional grid whose squares represent cells or ambient medium. A contact edge between two cells is assigned a lambda (λ) value ($\lambda \geqslant 0$), which represents the strength of adhesion. The sum (E) of the products of the number of contact edges and their respective λ values may be regarded as the negative or reciprocal of the total surface-free energy of the system. The maximally (or absolutely) stable structures (patterns) are those that have maximum E. We determined *all* the configurations with maximum E for two- and three-cell types, together with the constraints on the values of various λs. A method for extension to more than three types of cells was also given.

* First published in *J. Theor. Biol.* (1970) 28: 469–82.

For systems with two or three types of cells, none of the absolutely stable patterns is of significant histological importance, that is, none of these patterns is a two-dimensional analog of tubular, epithelial, or vesicular structures. However, we found that these structures for this system of isotropic nondeformable cells could be made locally stable by suitable choice of λs for a plausible set of motility rules. In this paper we examine the possibility of these analogs of histologically important structures as being absolutely stable patterns (ASPs) for systems of nondeformable *anisotropic* cells. Extending the method used in the preceding paper (Goel et al., 1970), we show in the next three sections that by assuming a very simple anisotropy, that is, by assuming that λ values for the edges of a square (cell) opposite each other are identical but different from those for the other two edges, analogs of above-mentioned structures are absolutely stable.

We first consider, in the next section, the case of two types of cells, one anisotropic, the other isotropic. This case includes the case of a system with only one type of anisotropic cells and a medium. We find *all* the ASPs and also give an algebraic method to prove that the set of ASPs that we find consists of all possible ASPs. We extend the determination of ASPs for a system consisting of three types of cells, one anisotropic and two isotropic. Here we find only some of the ASPs. For both these systems, some of the ASPs are found to be two-dimensional analogs of the biologically significant epithelial, tubular, and vesicular structures.

2. Absolutely Stable Patterns for Systems with Two Types of Cells, One Anisotropic, the Other Isotropic

We denote the types of cells by b and m and their numbers by N_b and N_m, respectively. Let the two opposite edges of a square (cell) be type 1 and the other two opposite edges be type 2. Let λ_{b1b1}, λ_{b1b2}, λ_{b2b2}, λ_{b1m}, λ_{b2m}, and λ_{mm} denote the λs for b_1b_1, b_1b_2, b_2b_2, b_1m, b_2m, and mm edges, respectively. For convenience in writing, we call the m cells type 3 and denote N_b and N_m by N_1 and N_3, respectively, and the above-mentioned six λs by λ_{11}, λ_{12}, λ_{22}, λ_{13}, λ_{23}, and λ_{33}, respectively.

As in Goel et al. (1970) let us assume that $N_1 \ll N_3$ so that we can neglect boundary effects that arise when the number of cells of any type at the boundary of the pattern changes from pattern to pattern. Further, as in the same paper, we make use of the fact that because the cells are nondeformable in our simple model, the number of edges of cells of each type are conserved. In other words, for any configuration,

$$2N_{11} + N_{12} + N_{13} = 2N_1 \tag{1a}$$

$$2N_{22} + N_{12} + N_{23} = 2N_1 \tag{1b}$$

$$2N_{33} + N_{13} + N_{23} = 4N_3. \tag{1c}$$

The E function of a pattern is

$$E = \lambda_{11}N_{11} + \lambda_{12}N_{12} + \lambda_{22}N_{22} + \lambda_{13}N_{13} + \lambda_{23}N_{23} + \lambda_{33}N_{33}. \quad (2)$$

Equations (1) and (2) are similar to the corresponding equations for the system with three types of isotropic cells and hence can be treated likewise. Using equations (1), the right-hand side of equation (2) can be expressed in terms of only three types of edges. There are $6!/3!3! = 20$ possible distinct groups of edges, each involving three types of edges. From these 20 groups, 3 groups $[(N_{11}, N_{12}, N_{13}), (N_{22}, N_{12}, N_{23}),$ and $(N_{33}, N_{13}, N_{23})]$ have to be discarded because, as can be seen from equations (1), E cannot be expressed in terms of Ns forming these groups. We wrote a FORTRAN program for use with an IBM 360 computer that will give the E in terms of the remaining 17 groups of N_{ij}'s. In table 1 we have given E in terms of the 17 groups

Table 1. Seventeen equivalent forms for E-function

Form	E-function forms
1	$N_{11}\mu_2 + N_{12}(\mu_2 + \mu_3 - \mu_1)/2 + N_{22}\mu_3$
2	$N_{11}\mu_2 + N_{12}(\mu_2 - \mu_1)/2 - N_{23}\mu_3/2$
3	$N_{11}(\mu_2 - \mu_3) + N_{12}(\mu_2 - \mu_1 - \mu_3)/2 + N_{33}\mu_3$
4	$N_{11}(\mu_1 - \mu_3) + N_{13}(\mu_1 - \mu_2 - \mu_3)/2 + N_{22}\mu_3$
5	$N_{11}\mu_1 + N_{13}(\mu_1 - \mu_2)/2 - N_{23}\mu_3/2$
6	$N_{11}\mu_1 + N_{13}(\mu_1 + \mu_3 - \mu_2)/2 + N_{33}\mu_3$
7	$N_{11}\mu_2 + N_{22}(\mu_1 - \mu_2) + N_{23}(\mu_1 - \mu_2 - \mu_3)/2$
8	$N_{11}(\mu_1 + \mu_2 - \mu_3)/2 + N_{22}(\mu_1 + \mu_3 - \mu_2)/2 + N_{33}(\mu_2 + \mu_3 - \mu_1)/2$
9	$N_{11}\mu_1 + N_{23}(\mu_2 - \mu_1 - \mu_3)/2 + N_{33}(\mu_2 - \mu_1)$
10	$N_{12}(\mu_3 - \mu_1)/2 - N_{13}\mu_2/2 + N_{22}\mu_3$
11	$-N_{12}\mu_1/2 - N_{13}\mu_2/2 - N_{23}\mu_3/2$
12	$-N_{12}\mu_1/2 + N_{13}(\mu_3 - \mu_2)/2 + N_{33}\mu_3$
13	$N_{12}(\mu_3 - \mu_1 - \mu_2)/2 + N_{22}(\mu_3 - \mu_2) + N_{33}\mu_2$
14	$-N_{12}\mu_1/2 + N_{23}(\mu_2 - \mu_3)/2 + N_{33}\mu_2$
15	$-N_{13}\mu_2/2 + N_{22}\mu_1 + N_{23}(\mu_1 - \mu_3)/2$
16	$N_{13}(\mu_3 - \mu_1 - \mu_2)/2 + N_{22}\mu_1 + N_{33}(\mu_3 - \mu_1)$
17	$N_{22}\mu_1 + N_{23}(\mu_1 + \mu_2 - \mu_3)/2 + N_{33}\mu_2$

obtained by using this program. For brevity we have left out the constant terms [independent of N_{ij} $(i, j = 1, 2, 3)$] because these terms are independent of pattern and hence are not important for the study of stability of patterns and depend on values of λs and N_1 and N_2. We have also used the notation

$$\mu_1 = \lambda_{11} + \lambda_{22} - 2\lambda_{12}$$

$$\mu_2 = \lambda_{11} + \lambda_{33} - 2\lambda_{13}$$

$$\mu_3 = \lambda_{22} + \lambda_{33} - 2\lambda_{23}.$$

The fact that we could express E in terms of three effective λs (which we call μs) instead of six λs is due to the conservation of three types of edges. In fact if there are all together r types of distinct edges, there will be $r(r + 1)/2$ different kind of boundaries (and hence λs) and r conservation relations. The number of effective λs (i.e., μs) will be $r(r + 1)/2 - r = r(r - 1)/2$, which is three for $r = 3$.

We now consider the absolutely stable patterns for various values of μs. If

$$\mu_i = 0, \qquad i = 1, 2, 3,$$

then by using any of the 17 equivalent forms, all the patterns are equally stable. On the other hand, if

$$\mu_1 = \mu_2 > 0, \qquad \mu_3 = 0,$$

the absolutely stable pattern has b cells clustered together so that N_{11} is maximum and m cells distributed outside the cluster in *any* form. The b cell cluster in which N_{11} is maximum is the one in which b cells are strung together in the form of a ring so that the edge of a type-1 b cell is in contact with edges of type-1 neighboring b cells. We will call this cluster b_1 ring and the b ring in which N_{22} is maximum b_2 ring. This b_2 ring will exist if $\mu_1 = \mu_3 > 0$, $\mu_1 = 0$. These rings are two-dimensional analogs of tubules and vesicles. If the number of b cells is not sufficient, b cells may not be able to form a ring without deforming themselves significantly. In that case, the structure with N_{11} (or N_{22}) maximum will just be a linear chain of b cells that is two-dimensional analog of a sheet or epithelium. For simplicity we will use the term *ring* both for closed and open rings.

To determine all absolutely stable patterns for the general case when two μs need not be equal, as in Goel et al., we have to consider each of the 17 forms of E separately and to find for what values of μ there exist one or more absolutely stable patterns. The procedure to determine all the absolutely stable patterns from one form has been described in Goel et al. For completeness we describe the procedure here also. The procedure is as follows: because E is maximum for an absolutely stable pattern, such a pattern must have the number of three types of edges occurring in a particular form either maximum or minimum. If we denote the maximum by 1 and the minimum by 0, we have $2^3 = 8$ possibilities for the number of the three types of edges, namely, 000, 001, 010, 100, 110, 101, 011, and 111. We investigate each of these eight cases to determine whether we can have a pattern with the number of edges of three types belonging to this case. If we do find such a pattern, it will be one of the absolutely stable ones. The corresponding values of μ can be found by setting the coefficient of the number of edges of a particular type less than or equal to 0 if this number is minimum and greater than 0 if it is maximum. This way we investigate all eight cases for each of the 17 forms.

We wrote a FORTRAN program for use with an IBM 360 computer that will investigate all the $17 \times 8 = 136$ cases and discard most of the cases for which there is no absolutely stable pattern (ASP). The logic used in the program to achieve this is that certain combinations of maximum or minimum number of edges were incompatible. For example, a pattern having maximum number of $b_1 b_1$ edges cannot also have maximum number of $b_1 m$ or $b_1 b_2$ edges. The program discards 77 cases; the remaining cases have to be examined manually. On doing so, it was found that 8 more cases do not lead to an ASP. Using this program we found that there are only 5 ASPs. In table 2 we have given the various ASPs together with the value of μs for these patterns in column 3. In column 1 we have given the form number (see table 1) that we have used and in column 2 the values (maximum or minimum) of number of edges of various types used to obtain the particular absolutely stable pattern. $b \times m$ stands for the pattern in which all the b cells are completely dispersed; $b_1 \times b_2$ stands for the checkerboard among b cells such that a type-1 edge is in contact with a type-2 edge, and $b_1 \times b_1$ stands for the checkerboard between b cells such that type-1 edge is in contact with another type-1 edge, a type-2 edge is in contact with another type-2 edge, and there is no contact between type-1 and type-2 edges.

We further determined the set of inequalities that define a region in μ space that covers the regions defined by all the other sets for the same pattern. This can most easily be done by an algebraic method that we now describe.

Table 2. Completely dispersed patterns

| Form | Patterns $b \times m$ | | | | | | Allowed μ's |
------	N_{11}	N_{12}	N_{22}	N_{13}	N_{23}	N_{33}	
1*	0	0	0				$\mu_2 \leq 0,\ \mu_2+\mu_3 \leq \mu_1,\ \mu_3 \leq 0$
2	0	0			1		$\mu_2 \leq 0,\ \mu_2 \leq \mu_1,\ \mu_3 < 0$
3	0	0				0	$\mu_2 \leq \mu_3,\ \mu_2 \leq \mu_1+\mu_3,\ \mu_3 \leq 0$
4	0		0	1			$\mu_1 \leq \mu_3,\ \mu_1 > \mu_2+\mu_3,\ \mu_3 \leq 0$
5	0			1	1		$\mu_1 \leq 0,\ \mu_1 > \mu_2,\ \mu_3 < 0$
6	0			1		0	$\mu_1 \leq 0,\ \mu_1+\mu_3 > \mu_2,\ \mu_3 \leq 0$
7	0		0		1		$\mu_2 \leq 0,\ \mu_1 \leq \mu_2,\ \mu_1 > \mu_2+\mu_3$
8	0		0			0	$\mu_1+\mu_2 \leq \mu_3,\ \mu_1+\mu_3 \leq \mu_2,\ \mu_2+\mu_3 \leq \mu_1$
9	0			1		0	$\mu_1 \leq 0,\ \mu_2 > \mu_1+\mu_3,\ \mu_2 \leq \mu_1$
10		0	0	1			$\mu_1 \geq \mu_3,\ \mu_2 < 0,\ \mu_3 \leq 0$
11		0		1	1		$\mu_1 \geq 0,\ \mu_2 < 0,\ \mu_3 < 0$
12		0		1		0	$\mu_1 \geq 0,\ \mu_3 > \mu_2,\ \mu_3 \leq 0$
13		0	0			0	$\mu_1+\mu_2 \geq \mu_3,\ \mu_2 \geq \mu_3,\ \mu_2 \leq 0$
14		0			1	0	$\mu_1 \geq 0,\ \mu_2 > \mu_3,\ \mu_2 \leq 0$
15			0	1	1		$\mu_2 < 0,\ \mu_1 \leq 0,\ \mu_1 > \mu_3$
16			0	1		0	$\mu_3 > \mu_1+\mu_2,\ \mu_1 \leq 0,\ \mu_1 \geq \mu_3$
17			0		1	0	$\mu_1 \leq 0,\ \mu_1+\mu_2 > \mu_3,\ \mu_2 \leq 0$

Table 2 (*continued*) Ring patterns

Form	N_{11}	N_{12}	N_{22}	N_{13}	N_{23}	N_{33}	Allowed μs.
		Pattern b_1 ring					
1	1	0	0				$\mu_2>0,\ \mu_2+\mu_3\leqslant\mu_1,\ \mu_3\leqslant0$
2	1	0			1		$\mu_2>0,\ \mu_2\leqslant\mu_1,\ \mu_3<0$
4	1		0	0			$\mu_1>\mu_3,\ \mu_1\leqslant\mu_2+\mu_3,\ \mu_3\leqslant0$
5	1			0	1		$\mu_1>0,\ \mu_1\leqslant\mu_2,\ \mu_3<0$
7	1		0		1		$\mu_2>0,\ \mu_1\leqslant\mu_2,\ \mu_1>\mu_2+\mu_3$
10*		0	0	0			$\mu_3\leqslant\mu_1,\ \mu_2\geqslant0,\ \mu_3\leqslant0$
11		0		0	1		$\mu_1\geqslant0,\ \mu_2\geqslant0,\ \mu_3<0$
15		0		0	1		$\mu_2\geqslant0,\ \mu_1\leqslant0,\ \mu_1>\mu_3$
		Pattern b_2 ring					
1	0	0	1				$\mu_2\leqslant0,\ \mu_2+\mu_3\leqslant\mu_1,\ \mu_3>0$
2*	0	0			0		$\mu_2\leqslant0,\ \mu_2\leqslant\mu_1,\ \mu_3\geqslant0$
4	0		1	1			$\mu_1\leqslant\mu_3,\ \mu_1>\mu_2+\mu_3,\ \mu_3>0$
5	0			1	0		$\mu_1\leqslant0,\ \mu_1>\mu_2,\ \mu_3\geqslant0$
7	0		1		0		$\mu_2\leqslant0,\ \mu_1>\mu_2,\ \mu_1\leqslant\mu_2+\mu_3$
10		0	1	1			$\mu_3\leqslant\mu_1,\ \mu_2<0,\ \mu_3>0$
11		0		1	0		$\mu_1\geqslant0,\ \mu_2<0,\ \mu_3\geqslant0$
15			1	1	0		$\mu_2<0,\ \mu_1>0,\ \mu_1\leqslant\mu_3$

Checkerboard patterns

Form	N_{11}	N_{12}	N_{22}	N_{13}	N_{23}	N_{33}	Allowed μ's
		Pattern $b_1\times b_2$					
1	0	1	0				$\mu_2\leqslant0,\ \mu_2+\mu_3>\mu_1,\ \mu_3\leqslant0$
2	0	1			0		$\mu_2\leqslant0,\ \mu_2>\mu_1,\ \mu_3\geqslant0$
3	0	1				1	$\mu_2\leqslant\mu_3,\ \mu_2>\mu_1+\mu_3,\ \mu_3>0$
4	0		0	0			$\mu_1\leqslant\mu_3,\ \mu_1\leqslant\mu_2+\mu_3,\ \mu_3\leqslant0$
6	0		0			1	$\mu_1\leqslant0,\ \mu_1+\mu_3\leqslant\mu_2,\ \mu_3>0$
7	0		0		0		$\mu_2\leqslant0,\ \mu_1\leqslant\mu_2,\ \mu_1\leqslant\mu_2+\mu_3$
8	0		0			1	$\mu_1+\mu_2\leqslant\mu_3,\ \mu_1+\mu_3\leqslant\mu_2,\ \mu_2+\mu_3>\mu_1$
9	0				0	1	$\mu_1\leqslant0,\ \mu_2\leqslant\mu_1+\mu_3,\ \mu_2>\mu_1$
10		1	0	0			$\mu_3>\mu_1,\ \mu_2\geqslant0,\ \mu_3\leqslant0$
11		1	0	0			$\mu_1<0,\ \mu_2\geqslant0,\ \mu_3\geqslant0$
12		1	0			1	$\mu_1<0,\ \mu_3\leqslant\mu_2,\ \mu_3>0$
13		1	0			1	$\mu_3>\mu_1+\mu_2,\ \mu_3\leqslant\mu_2,\ \mu_2>0$
14		1			0	1	$\mu_1<0,\ \mu_2\leqslant\mu_3,\ \mu_2>0$
16			0	0		1	$\mu_3\leqslant\mu_1+\mu_2,\ \mu_1\leqslant0,\ \mu_3>\mu_1$
17			0		0	1	$\mu_1\leqslant0,\ \mu_1+\mu_2\leqslant\mu_3,\ \mu_2>0$
		Pattern $b_1\times b_1$					
11*		0		0	0		$\mu_1\geqslant0,\ \mu_2\geqslant0,\ \mu_3\geqslant0$
12		0		0		1	$\mu_1\geqslant0,\ \mu_3\leqslant\mu_2,\ \mu_3>0$
14		0			0	1	$\mu_1\geqslant0,\ \mu_2\leqslant\mu_3,\ \mu_2>0$

The basis for the method is that if we know that there are two regions in μ space that are adjacent to each other, the set that defines the combined region will consist of all the inequalities of the two sets defining the adjacent regions. On physical grounds we know that all the sets of inequalities for a particular pattern define regions so that every region has at least one region adjacent to it. In other words, these sets define a continuous region. If one set has an inequality $\alpha > 0$ and another set has $\alpha < 0$, these two sets define regions adjacent to each other. We select pairs of such sets and combine them into one set per pair. We then select pairs from this reduced number of sets and combine them.

We continue the process until we cannot combine any further. The resulting set of inequalities will define the region that defines a set covering the regions defined by all the sets for the pattern under consideration. It may be noted that this procedure is equivalent to combining all the sets defining a particular pattern without checking if the regions are next to one another. For the patterns $b \times m$, b_1 ring, b_2 ring, and $b_1 \times b_1$ checkerboard, the resulting sets are the ones with an asterisk near the form number in table 2. For the $b_1 \times b_2$ checkerboard pattern, the resulting set is

$$\mu_1 \leqslant 0, \qquad \mu_1 \leqslant \mu_3, \qquad \mu_1 \leqslant \mu_2, \qquad \mu_1 \leqslant \mu_2 + \mu_3.$$

It may be noted that the inequalities of the set that define the region covering all other regions defined by the other sets of inequalities for the same ASP will also be true for all other sets of inequalities for the same pattern.

To determine whether the patterns described above are the only ASPs, we have to ascertain whether the values of μ given in column 3 of table 2 cover the whole μ_1, μ_2, μ_3 space. This can be determined either by the geometric method described in Goel et al. or by an algebraic method that we will describe in the next section. In the geometric method we construct a polyhedron in μ_1, μ_2, μ_3 space so that its various faces define the regions and give various ASPs. If the polyhedron is a closed one, the values of μ given in column 3 of table 2 cover the whole μ-space and the patterns described above are the only ASPs. When we construct the polyhedron, it is found to be a closed one and thus we have all the ASPs. As mentioned in Goel et al., this geometrical method will fail for systems with four or more types of isotropic cells or, in the present case of anisotropic cells, if we have one or more additional types of cells because in either case the space will be six or more dimensional. The algebraic method that we will describe is applicable for μ space of any dimension.

Having found all the ASPs for the system consisting of one type of anisotropic cells, let us take a look at the ASPs. There are three new patterns that do not exist for the system with isotropic cells. These patterns are b_1 ring, b_2 ring, and checkerboard with $b_1 b_2$ minimum. b_1 and b_2 rings are the two-dimensional analogs of vesicles and closed shells; an open b_1 or b_2 ring is the analog of sheets. We may note that these analogs were not absolutely stable structures for the isotropic cells (Goel et al., 1970). Checkerboard with $b_1 b_2$ minimum does not have any biologically interesting three-dimensional analog.

3. ALGEBRAIC METHOD TO FIND THE EXTENT OF μ SPACE COVERED BY A SET OF INEQUALITIES

We will describe the method by taking the example of system with two types of cells, one isotropic and the other anisotropic, investigated in the previous section. There are 17 forms of the E function. For each form, by putting the coefficients of $3N_{ij}$'s less than zero and greater than zero, one gets eight sets of inequalities, each set consisting of the subsets of inequalities among μs. The eight sets of inequalities for any of the 17 forms of E function will cover the whole μ-space. Each of the eight regions defined by these eight sets of inequalities may be covered by either a single absolutely stable pattern (ASP) or more than one ASP, or may be completely or partially empty. By the method of finding ASPs, one can easily find the regions that are covered by a single ASP. We choose the form of E for which the maximum number of sets of inequalities define regions covered by a single ASP. For example, for the case of system under consideration, we choose form 1. The eight sets of inequalities are

$$\mu_1 > 0 \qquad \mu_2 > 0 \qquad \mu_3 > 0 \qquad\qquad \text{(i)}$$

$$\mu_1 > 0 \qquad \mu_2 > 0 \qquad \mu_3 < 0 \qquad\qquad \text{(ii)}$$

$$\mu_1 > 0 \qquad \mu_2 < 0 \qquad \mu_3 > 0 \qquad\qquad \text{(iii)}$$

$$\mu_1 > 0 \qquad \mu_2 < 0 \qquad \mu_3 < 0 \qquad\qquad \text{(iv)}$$

$$\mu_1 < 0 \qquad \mu_2 > 0 \qquad \mu_3 > 0 \qquad\qquad \text{(v)}$$

$$\mu_1 < 0 \qquad \mu_2 > 0 \qquad \mu_3 < 0 \qquad\qquad \text{(vi)}$$

$$\mu_1 < 0 \qquad \mu_2 < 0 \qquad \mu_3 > 0 \qquad\qquad \text{(vii)}$$

$$\mu_1 < 0 \qquad \mu_2 < 0 \qquad \mu_3 < 0 \qquad\qquad \text{(viii)}$$

As can be seen from table 2, sets (i)–(v) define regions covered by the pattern $b_1 b_1$ checkerboard, b_1 ring, b_2 ring, $b \times E$, and $b_1 b_2$ checkerboard, respectively. We will now have to examine regions (vi) to (viii) to see if they are covered by ASPs or are completely or partially empty.

To find the answer we search in table 2 for two or more sets of inequalities for neighboring regions of μ space that, when combined by the method discussed in section 2, cover one of the missing regions (vi)–(viii). To cover region (vi), we can find three such sets of inequalities:

(1) $\mu_1 < 0$ $\mu_1 + \mu_2 < \mu_3$ $\mu_2 > 0$ from $b_1 b_2$ checkerboard

(2) $\mu_1 < 0$ $\mu_1 + \mu_2 > \mu_3$ $\mu_3 > \mu_1$ from $b_1 b_2$ checkerboard

(3) $\mu_1 < 0$ $\mu_2 > 0$ $\mu_1 > \mu_3$ from b_1 ring.

Regions (1) and (3) are both neighbors of region 2 because they share one boundary and are on opposite sides of a second boundary. If all three sets are combined, the resulting set is

$$\mu_1 < 0 \qquad \mu_1 > 0.$$

This set obviously contains the more restricted region (vi). By similar reasoning the following three sets of inequalities:

(1) $\mu_1 < 0$ $\mu_2 > \mu_1 + \mu_3$ $\mu_3 > 0$ from $b_1 b_2$ checkerboard

(2) $\mu_1 < 0$ $\mu_2 < \mu_1 + \mu_3$ $\mu_2 > \mu_1$ from $b_1 b_2$ checkerboard

(3) $\mu_1 < 0$ $\mu_3 > 0$ $\mu_1 > \mu_2$ from b_2 ring

can be combined to give the set

$$\mu_1 < 0, \qquad \mu_3 > 0,$$

which contains the more restricted region (vii).

To get set (viii), the following two sets of inequalities are combined:

(1) $\mu_2 < 0$ $\mu_2 + \mu_3 > \mu_1$ $\mu_3 < 0$ from $b_1 b_2$ checkerboard

(2) $\mu_2 < 0$ $\mu_2 + \mu_3 < \mu_1$ $\mu_3 < 0$ from $b \times m$

to give the set

$$\mu_2 < 0 \qquad \mu_3 < 0,$$

which obviously contains within it region (viii).

We have now shown that regions (vi)–(viii), as well as regions (i)–(v), are contained in ASPs listed in table 2. This result proves that all of the three-dimensional μ space is covered by the five patterns in table 2 and that there are no more ASPs.

4. Absolutely Stable Patterns for Systems with Three Types of Cells, One Anisotropic and Two Isotropic

We now extend the procedure of section 2 to the system consisting of three types of cells with only one type of anisotropic cell. This method also applies to the system with one type of anisotropic cell, one type of isotropic cell, and medium. Let us denote the three types of cells by b, w, and m, with b as the anisotropic type with two types of edges, b_1 and b_2. There are thus ten types

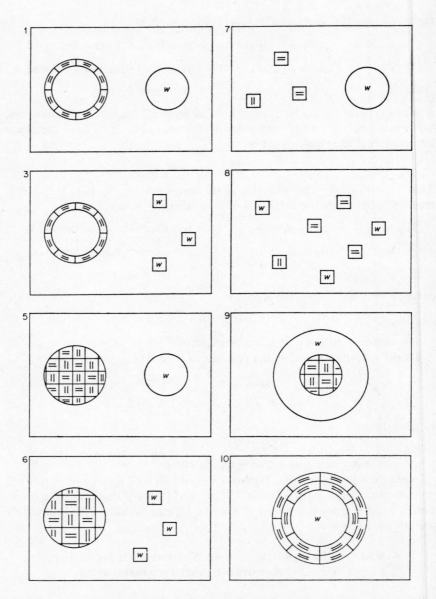

Fig. 1. Some of the absolutely stable patterns for a system with one anisotropic and two isotropic types of cells. The patterns obtained by interchanging b_1 and b_2 edges are omitted. $=$ has been used to indicate the orientation of the b cells, b_2 edge is parallel to $=$.

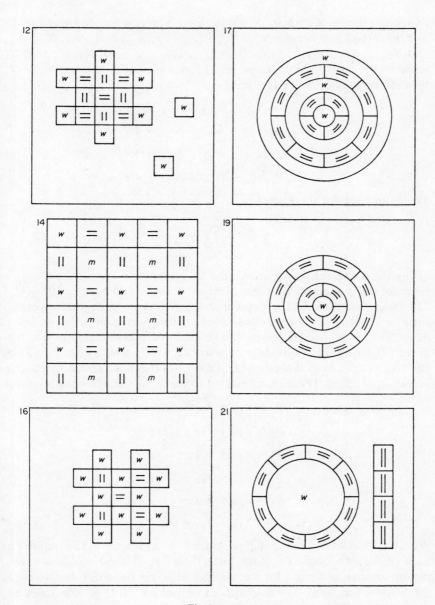

Fig. 1—cont.

of possible edges: $b_1b_1, b_1b_2, b_1w, b_1m, b_2b_2, b_2w, b_2m, ww, wm$, and mm. Let us denote their numbers by $N_{11}, N_{12}, N_{13}, N_{14}, N_{22}, N_{23}, N_{24}, N_{33}, N_{34}$, and N_{44}, respectively, and their λs by $\lambda_{11}, \lambda_{12}, \ldots, \lambda_{44}$. If N_1, N_3, and N_4, respectively, denote the total number of b, w, and m cells, and $N_1, N_3 \ll N_4$, then the conservation of edges implies

$$2N_{11} + N_{12} + N_{13} + N_{14} = 2N_1$$

$$2N_{22} + N_{12} + N_{23} + N_{24} = 2N_1$$

$$2N_{33} + N_{13} + N_{23} + N_{34} = 4N_3$$

$$2N_{44} + N_{14} + N_{24} + N_{34} = 4N_4.$$

The E function for the system is

$$E = \sum_{\substack{i,j=1 \\ i<j}}^{4} \lambda_{ij} N_{ij}.$$

Using the conservation equations, the E function can be expressed only in terms of the number of six types of edges. There are altogether $10!/6!4! = 210$ groups of edges, each group consisting of 6 types of edges. However, it turns out that the E function cannot be expressed in terms of 69 groups of edges. The remaining 141 groups can be used to express the E function. Each of the 141 forms of E will involve only six effective λs, which we call μs. These effective λs arise from those combinations of λs that are allowed by the conservation of edges. These μs are related to the λs by the following relations:

$$\mu_1 = \lambda_{11} + \lambda_{22} - 2\lambda_{12}$$

$$\mu_2 = \lambda_{11} + \lambda_{33} - 2\lambda_{13}$$

$$\mu_3 = \lambda_{11} + \lambda_{44} - 2\lambda_{14}$$

$$\mu_4 = \lambda_{22} + \lambda_{33} - 2\lambda_{23}$$

$$\mu_5 = \lambda_{22} + \lambda_{44} - 2\lambda_{24}$$

$$\mu_6 = \lambda_{33} + \lambda_{44} - 2\lambda_{34}.$$

We wrote a FORTRAN computer program to examine the ASPs implied by each of the 141 forms. For each form there are $2^6 = 64$ combinations of maximum and minimum for the number or edges occurring in any of the forms for E functions. The computer examined $64 \times 141 = 9024$ cases and discarded most of the combinations that were incompatible in the sense described in section 2. The program gave 1029 combinations (which may lead to the ASP) that had to be examined manually to see if any of the combinations were incompatible. Out of 1029, only 458 combinations were found to be compatible; these combinations gave only 22 ASPs (fig. 1). The conditions

Table 3. Conditions on μ's for the absolutely stable patterns for a system with one anisotropic and two isotropic types of cells

Pattern no.	Pattern	Allowed μ's
1	b_1 ring, ⓦ	$5\leq$ $3\geq$ $6>$ $2-6\geq$ $1-5>$ $4-5-6\geq$
2	b_2 ring, ⓦ	$3\leq$ $5\geq$ $6>$ $1-3\geq$ $4-6>$ $2-3-6\geq$
3	b_1 ring, $w\times m$	$5\leq$ $3\geq$ $6\leq$ $1-5>$ $2-6\geq$ $4-5-6\geq$
4	b_2 ring, $w\times m$	$3\leq$ $5\geq$ $6\leq$ $1-3>$ $4-6>$ $2-3-6\geq$
5	$b_1\times b_2$, ⓦ	$3\leq$ $5\geq$ $6\geq$ $1-3-5<$ $2-3-6\geq$ $4-5-6\geq$
6	$b_1\times b_2$, $w\times m$	$3\leq$ $5\geq$ $6\leq$ $1-3-5<$ $2-3-6\geq$ $4-5-6\geq$
7	$b\times m$, ⓦ	$3\leq$ $5\geq$ $6\geq$ $1-3\geq$ $2-3-6\geq$ $4-5-6\geq$
8	$b\times w\times m$	$3\leq$ $5\geq$ $6\leq$ $1-3\geq$ $2-3-6\geq$ $4-5-6\geq$
9	$b_1\times b_2$ covered by w	$2\leq$ $4\leq$ $6\geq$ $1-2-4<$ $2-3+6\geq$ $4-5+6\leq$
10	ⓦ covered by b_1 rings	$1\geq$ $4\geq$ $5\geq$ $2-4\geq$ $3-5>$ $4+5+6\leq$
11	ⓦ covered by b_2 rings	$1\geq$ $2\geq$ $3\geq$ $2-4\leq$ $3-5>$ $2+3-6\leq$
12	$b_1\times b_2$, $w\times m$ no b_1m	$5\leq$ $6\geq$ $2-6\leq$ $2-3-6\geq$ $4-5-6\geq$ $1-2-5+6<$
13	$b_1\times b_2$, $w\times m$ no b_2m	$3\leq$ $6\geq$ $4-6\leq$ $2-3-6\geq$ $4-5-6\leq$ $1-3-4+6<$
14	$b_1\times w$, $b_2\times m$	$2\leq$ $5\geq$ $2-3-6\geq$ $2-4+5\leq$ $1-2-5+6>$ $2-3-4+5\leq$ $1-2+4-(2)5>$
15	$b_2\times w$, $b_1\times m$	$4\geq$ $3\leq$ $4-5-6\geq$ $1+3-4<$ $2-3-4+6>$ $2-3-4+5>$ $1+2-(2)3-4>$
16	$b\times w$	$2\leq$ $5>$ $4\leq$ $2-6\leq$ $4-5-6\leq$ $2-3+6\geq$ $1-2+4-(2)5\leq$; $4-6>$ $4-5+6\leq$ $2-3-6\geq$ $1-2-4+(2)6>$
17	Multilayered b_1 onion, no b_1m	$2\geq$ $4\leq$ $6>$ $1-4>$ $3-6\geq$ $4-5+6\leq$
18	Multilayered b_2 onion, no b_2m	$2\leq$ $4\geq$ $6>$ $1-2\geq$ $5-6\geq$ $2-3+6\geq$
19	Multilayered b_1 onion, no wm	$2\geq$ $4\leq$ $1-4\geq$ $4-5<$ $4-5+6\geq$ $3+4-5>$
20	Multilayered b_2 onion, no wm	$2\leq$ $1-2\geq$ $2-3<$ $2-3+5\geq$ $2-3+6\geq$
21	ⓦ covered by single b_1 ring	$3\geq$ $5\leq$ $1-5\geq$ $4-5\geq$ $2-4+5\geq$ $4-5-6<$
22	ⓦ covered by single b_2 ring	$3\leq$ $5\geq$ $1-3\geq$ $2-3>$ $2-3-6\geq$ $2-3-4\geq$

For simplicity, only the subscripts of μ's are given and the zeros on the right hand of the sign of inequality have been omitted. Thus $5<$ means $\mu_5<0$.

on μs for each pattern were combined by using the method discussed in section 2 into one set of conditions. This set will define the region in μ space for a particular pattern. These sets of inequalities for various ASPs are given in table 3. An analysis of the total six-dimensional μ space by the method discussed in section 3 shows that there are some regions in the μ space that are not covered by the 22 ASPs given in figure 1. These missing regions correspond to patterns transitional between 2 ASPs.

An examination of the ASPs reveals that there are some patterns that are two-dimensional analogs of the biologically interesting structures. For example, patterns 21 and 22 (which is the same as 21 except for the interchange of b_1 and b_2 edges) are two-dimensional analogs of an epithelium, that is, a mass of one type of cells covered by a single layer of another, or a single layer tubular structure with inside and outside consisting of different type of cells. The analysis shows that excess of anisotropic cells will form a line or ring, which are two-dimensional analogs of a sheet and vesicular structure, respectively.

This work was started by one of us (N. S. G.) while attending the Colloquium on Theoretical Biology in Fort Collins in July and August 1969, which was arranged by the American Institute of Biological Sciences and supported by the National Aeronautics and Space Administration. One of us (N. S. G) was supported in part by the U.S. Office of Naval Research. The other (A. G. L.) was supported in part by the National Institutes of Health Training Grant to the Department of Biology, University of Rochester. We gratefully acknowledge the use of the facilities of the University of Rochester Computing Center.

REFERENCES

Goel, N., Campbell, R. D., Gordon, R., Rosen, R., Martinez, H., and Yčas, M. (1970) *J. Theor. Biol.* 28: 423.
Steinberg, M. S. (1963) *Science, N.Y.* 141: 401.
Steinberg, M. S. (1964) In *Cellular Membranes in Development*. ed. M. Locke. New York: Academic Press.

9. Simulation of Movement of Cells during Self-Sorting

Ardean G. Leith and Narendra S. Goel*

Using a model proposed earlier by Goel et al. and a set of plausible motility rules, a computer simulation of self-sorting of two types of randomly mixed cells is carried out. It is shown that for sorting to the expected final structure, the important aspects of the motility rules are (1) the distance up to which a cell can "sense" the presence of its neighboring cells and then move, and (2) the lack of decrease of a certain E function (which is a measure of total adhesion). The final structures, which are given for a set of rules bear a good resemblance of these structures to those experimentally sorted out.

1. Introduction

The wide variety of phenomena that lead to some understanding of morphogenesis include the self-sorting-out of cells that occurs when many types of cells or tissues are mixed. This self-sorting has been studied both experimentally (Steinberg, 1970, and the references cited therein; Trinkaus, 1969) and theoretically (Steinberg, 1970; Goel et al., 1970; Goel and Leith, 1970). Briefly, in general, when two or more tissues are disaggregated by appropriate change in the environmental conditions (e.g., pH, concentration of Ca^{2+}, and so forth) and the resulting cells are mixed, the cells sort out to form certain structures. Further, the structure formed depends on the types of cells and is path independent in that the same structure is formed from randomly interdispersed cells that is formed from the fusion of the intact fragments of tissues. On the basis of the experiments on self-sorting of several types of tissue, taken two and more at a time, Steinberg proposed that two properties, motility and differential adhesion, are sufficient to account for cell sorting and to determine the final structure.

However, Steinberg's hypothesis leaves open the question of the motility rules (i.e., the factors that describe the movement) that cells may follow. In this paper we attempt to throw some light on these motility rules.

The model used for analysis is the one proposed earlier (Goel et al., 1970). The system was modeled in two dimensions by a space tessellated or subdivided into equal squares. Each square can be occupied by a cell of any type

* First published in *J. Theor. Biol.* (1971) 33: 171–88.

including medium, which for convenience is treated as another cell type. Each edge of contact between two cells is assigned an affinity λ, which is a measure of the strength of binding between the cell types. Thus, λ_{bb} and λ_{bw} are the affinities between cells of type b and b and between b and w, respectively. We also defined an E function for a pattern of cells as the sum of the number of edges multiplied by their λ values. For a system with two types of cells (b and w),

$$E = \lambda_{bb}N_{bb} + \lambda_{bw}N_{bw} + \lambda_{ww}N_{ww}, \tag{1}$$

where N_{bb} is the number of bb edges, and so forth. The pattern of maximum stability (called an absolutely stable pattern) is the one that has maximum E.

Using the above-mentioned model, we explicitly calculated the absolutely stable patterns for both the isotropic (Goel et al., 1970) and anisotropic cells (Goel and Leith, 1970) for systems with two and three types of cells, and we gave the method to calculate absolutely stable patterns for a system with more than three types of cells. For a system with two isotropic cell types, it was shown that if

$$\lambda_{bb} > \tfrac{1}{2}(\lambda_{bb} + \lambda_{ww}) > \lambda_{bw} > \lambda_{ww}, \tag{2}$$

the absolutely stable pattern is the one for which N_{bb} and N_{ww} are at a minimum, that is, the pattern in which b cells are closely packed and surrounded by w cells. Ths pattern was called the onion pattern. For other values of λ, other patterns are absolutely stable.

The basic method we used to determine the motility rules was also put forth in Goel et al. (1970). In this method, for simplicity, the case of $\lambda_{bb} = 1$, $\lambda_{ww} = \lambda_{bw} = 0$ was chosen so that from condition (2) the absolutely stable pattern is the onion pattern. The aggregation of randomly distributed cells of types b and w was simulated by using a computer and a set of plausible motility rules. Specifically in this simulation, a certain fraction (f_b) of a 20×20 square grid was filled in random positions by b cells, and the rest of the grid was considered to be occupied by w cells. This arrangement resulted in each cell touching four cells, except for those cells at the surface of the grid. In order to simplify the problem, the cells were confined to the grid with a fixed shape. This is analogous to the experimental situation in which one has a randomly mixed aggregate and the cells do not migrate outward into the medium from it. The b cells were allowed to move from one place to another by exchanging two cells using some plausible motility rules, and we asked whether an onion pattern can be reached from a random distribution of cells for the above values of λs. The postulated motility rules will be inadequate if they do not lead to such a result. None of these rules, which were used in Goel et al. (1970), which we will explicitly give in the next section, and which at first seemed quite reasonable, gave the onion pattern. There was a considerable amount of trapping of *w* cells in the *b* cluster regardless of the proportions of two-cell types. The three-dimensional grid gave basically the same results as the two-dimensional grid.

In this paper we make a systematic study of a large variety of plausible motility rules for cell aggregation on a two-dimensional grid. No claim is made that all the motility rules have been exhausted. Instead, the motility rules studied are such that the effects of the following on the aggregation can be determined.

(a) Change in the selection of the cell that moves
(b) Changes in the direction and "distance" that a cell can move
(c) Allowing both *b* and *w* cells to move rather than only *b* cells
(d) Giving priority to moves in the same direction that a cell has previously moved
(e) Scanning all the cells to find which *b* cell movement will lead to maximum N_{bb}, and selecting this move or, if there is more than one movement leading to the same N_{bb}, selecting one of these movements randomly
(f) Giving priority to moves which resulted in larger aggregates, that is, including cooperative effects
(g) Increasing sorting time by letting the program run for a longer time
(h) Changes in the shape of the cell by changing the grid (e.g., hexagonal grid)

The effects of these general features are discussed in the next section.

2. MOTILITY RULES AND RESULTS

In this section we describe the various motility rules we have studied and give the results obtained. With few exceptions, as in Goel et al. (1970), for all the rules studied, we use a random number generator to place *b* cells on a 20×20 array in a FORTRAN program at random with probability f_b, with the rest of the array containing *w* cells. Thus f_b is the fraction of *b* cells. This simulates the random distribution of *b* and *w* cells. The exceptions to the above distribution are either when we simulate in three dimensions or take the grid to be hexagonal. In Goel et al. (1970), a strip of *w* cells, two wide around the array, was used to confine the *b* cells to the grid. Because in the present study we 2ave a larger variety of rules, this border is not enough to confine the *b* cells to the grid. The same result is achieved by assuming the interaction of *b* and *w* cells with the border the same and by not allowing either type to move beyond the border. The details of the scanning and movement of the cells are part of the motility rule. We first describe the three rules described by Goel et al. and then describe the other rules studied in this paper.

The three rules are as follows.

Rule (a)

All the b cells are scanned one at a time during each computing cycle (or time step) but in random order to avoid systematic effects. The increase in nearest neighbor b–b pairs, that is, ΔN_{bb}, which would occur when a b cell moves to any of its eight nearest neighbor locations, is calculated. The cell is moved to the neighboring location for which ΔN_{bb} is maximum but ≥ 0. If, for two or more locations, ΔN_{bb} is maximal, one of them is chosen at random. If for none of the eight nearest neighbor locations $\Delta N_{bb} \geq 0$, the move is not made. The program then chooses another b cell. Every cell is given an opportunity to move before a cell gets another chance to move. The computing cycle is stopped after each b cell has 100 opportunities to move or after a scan produces no further movement.

Rule (b)

This rule is exactly the same as (a) except that b cells are allowed to move if $\Delta N_{bb} \geq 1$.

Rule (c)

Here the rule is the same as (a) except that b cells are allowed to move if $\Delta N_{bb} \geq -1$.

The additional motility rules we have studied in this paper are as follows.

Rule 1

This rule is the same as rule (a) mentioned above except that cell movements are allowed only to four *nondiagonal* nearest neighbor spaces. As in rule (a), a cell is chosen randomly to move but is not allowed to move until all the other cells have had an opportunity to move. ΔN_{bb} is ≥ 0 and the move in which ΔN_{bb} is greatest is chosen. If there is more than one such move, one is chosen randomly.

Rule 2

This rule is the same as rule 1 except that a cell moves to only one of the four diagonal nearest neighbor spaces. This rule and rule 1 are unrealistic but, together with rule (a), they give an idea of the importance of diagonal moves.

Rule 3

In this rule, all eight (four diagonal and four nondiagonal) nearest neighbor moves are allowed provided $\Delta N_{bb} \geq 0$, the same as in rule (a). However, in this rule, though the direction in which ΔN_{bb} is maximum is chosen, if there is more than one direction in which ΔN_{bb} is maximum, the direction in which the cell moved in the previous scan is chosen rather than a random direction. This system simulates movement in which a cell has a higher than random probability to continue moving in the same direction in which it once started moving.

Rule 4

This rule retains all the features of rule 3 except that of the scanning procedure. A cell is no longer unable to move until all the other cells have had an opportunity. Instead, in the random selection procedure, a cell can be chosen to move any number of times before other cells have an opportunity to move. This method simulates the system in which every cell can move any time, but there may be cells that never move.

Rule 5

As in rules (a), 3, and 4, this rule only allows moves to any of the eight neighboring spaces and with $\Delta N_{bb} \geqslant 0$. However, in addition, any cell that completes a move to a neighboring space having *no bb* contact, or a move that increases *bb* contact, is eligible for movement before other cells that did not complete such a move. Also, as in rule 3, the moves in the same direction as the previous move are preferred. Allowing cells that moved to a space having no *bb* contact to move again speeds up the sorting. In this simulation of activated movement a move that results in increased *bb* adhesion (increase in E) makes that cell more likely to continue moving and gives preference to the direction of such a move.

Rule 6

This rule is exactly the same as rule (a) except that here both *b* and *w* cells are moved. In this rule, as in (a), a cell does not move until all the other cells have an opportunity to move. The move is allowed to any of the eight neighboring spaces with $\Delta N_{bb} \geqslant 0$, and the move with maximum ΔN_{bb} is preferred. In addition, *w* cells also actively move to one of the eight neighboring places for which $\Delta N_{ww} \geqslant 0$, and the move with maximum ΔN_{ww} is preferred. This movement of *w* cells also either keeps N_{bb} constant or increases it. The simulation with this rule perhaps has a higher correspondence to reality in that both types of cells are assumed to have similar types of motility.

Rule 7

This rule is quite different from the above rules in that we only require that $\Delta N_{bb} \geqslant 0$ and not necessarily the largest ΔN_{bb}. A cell moves randomly to any of the eight neighboring spaces as long as $\Delta N_{bb} \geqslant 0$. This rule simulates a system in which the force driving the aggregation is the lack of decrease of E function rather than the maximization of the E function at each movement.

Rule 8

This rule is similar to rule (a) except that the selection of the cell for movement is not random. Here, after the b cells have been placed randomly on the grid, ΔN_{bb} is calculated for all (eight) possible moves of *each* cell. The cell whose movement increases the bb contact most is chosen to move. If there is more than one such cell, another is chosen randomly. The procedure for the calculation of N_{bb} for all the cells is again repeated, and then again the cell with maximum N_{bb} is chosen. This procedure is continued until either no further moves are possible or, for a number of moves that is roughly one and a half times the number of moves required by other rules for complete simulated aggregation, whichever occurs first. The latter criterion was arbitrary and was dictated by the cost for the simulation. The simulation based on this rule resembles a system in which the direction and occurrence of a movement of a cell are very strongly affected by the increase in E function (i.e., increase in adhesion).

Rule 9

In this rule the cells are scanned as in rule (a) in that moves are made only to the eight neighboring locations so that $\Delta N_{bb} \geqslant 0$. The move for which ΔN_{bb} is maximum is chosen. However, if there is more than one such move, the move that takes the cell to the largest cluster is selected. If there are two clusters of equal size, one is chosen at random. This rule simulates a system with cooperative effects, that is, the larger the number of cells in a cluster, the higher the probability of a b cell becoming attached to it.

Rule 10

In this rule each b cell is assumed to have an ability to feel the presence of other cells (and to move to locations) separated from it by two cell diameters. Thus each cell is allowed to move to 8 nearest-neighbor and 12 next-nearest-neighboring spaces (excluding the next-nearest-neighboring diagonal spaces). The movement is made as in (a) except that a move to the next-nearest-neighboring location is only allowed if a move to the nearest neighbor location is not possible (i.e., does not satisfy the condition $\Delta N_{bb} \geqslant 0$).

Rule 11

This rule is the same as rule 10 except that no preference is given to move to the nearest neighbor locations. In other words, all 20 neighboring spaces are considered equivalent and the move is made to the location for which ΔN_{bb} ($\geqslant 0$) is maximum.

Rule 12

In this rule each b cell is assumed to have an ability to feel the presence of other cells (and to move to locations) separated from it by three cell diameters.

Each cell is allowed to move to 8 nearest neighboring, 16 next-nearest-neighboring, and 12 next-to-next nearest neighboring spaces (these 12 spaces do not include the four next-to-next nearest neighbor diagonals and 8 next to the diagonal spaces). The cell can move to any of these 36 neighboring positions so as to maximize ΔN_{bb} but will move only if $\Delta N_{bb} \geq 0$. Thus, this rule is the same as rule (a) except a cell can "feel" up to three layers of surrounding cells and move to locations separated from it by three cell diameters.

Rule 13

This pattern is similar to rule 5 in that a cell that has completed a move to a space that increases *bb* contact, or that has no *bb* contact, can move again before other cells that did not complete such a move. The cell can move to any of the 36 neighboring locations, all considered equivalent. Again, the move with maximum ΔN_{bb} is chosen.

Rule 14

In this rule also, all 36 neighboring locations are considered equivalent, and the move with maximum ΔN_{bb} is chosen. However, if there is more than one move with maximum ΔN_{bb}, the move in the direction in which the cell moved when last scanned is preferred. Further, as in rule 5, any cell that completes a move to a neighboring space having no *bb* contact or a move that increases *bb* contact is eligible for movement before other cells that did not complete such a move.

Rule 15

This rule is similar to rule 12 except that the move to next-nearest neighbor location is allowed only if a move to the nearest neighbor location is not possible. Likewise, the move to the next-to-next nearest neighbor location is only made if a move to the next-nearest neighbor location is not possible. Of course, ΔN_{bb} must always be ≥ 0 for a move to occur.

Rule 16

Here again the cell can move to any one of the 36 neighboring spaces, and nearest and next-nearest neighbor locations are given preference, as in rule 15, and the eligibility of a cell that has moved for further moves, the feature of rule 13, is retained.

Rule 17

This rule is rule 16 with the additional feature that, if there are two locations with maximum ΔN_{bb}, the move in the direction in which the cell moved when last scanned is preferred.

Rule 18

In this, rule the cell can move to any of the 36 neighboring spaces, all considered equivalent but without the restriction of only maximum N_{bb} moves. The cell to be moved is randomly chosen as long as $\Delta N_{bb} \geqslant 0$. As in rule (a), every cell is given a chance to move before a cell is allowed to move again.

Rule 19

In this rule both b and w cells are moved to any of the 36 nearest spaces so as to maximize ΔN_{bb} (with ΔN_{bb} or $\Delta N_{ww} \geqslant 0$), and moves to the inner layers are given preference. The scanning of cells feature of rule (a) is retained.

Rule 20

This system is the same as rule 19 except that all moves to all the 36 spaces are considered equivalent.

Rule 21

This rule is the same as rule 20 with the addition of preference to move in the same direction as a cell previously moved and of the "rescanning of cells that completed successful moves" feature of rule 13.

Rule 22

In this rule the cells are allowed to move to 44 spaces. These 44 spaces include the 36 neighboring spaces described in rule 12. In addition, 8 next to the diagonal of the next-to-next nearest neighboring layer are included. Other than that, the rule is the same as rule 18, that is, the cells can move to any of these 44 neighboring spaces and all are considered equivalent but without the restriction of only maximum ΔN_{bb} moves. The cell to be moved is randomly chosen as long as $\Delta N_{bb} \geqslant 0$. Further, as in rule (a), every cell is given a chance to move before a cell is allowed to move again.

Rule (i)

This rule is the same as (a) except that the cells are taken to be hexagonal. Thus each cell is allowed to move to any of the six nearest neighbor spaces. The cell is moved to that neighboring location for which ΔN_{bb} is maximum and $\geqslant 0$. If for two or more locations ΔN_{bb} is maximal, one of them is chosen at random. Every cell is given an opportunity to move before a cell gets another chance to move.

Rule (ii)

In this rule the hexagonal cells are allowed to move to any of the 18 neighboring locations (6 nearest and 12 next-nearest neighbors), all locations considered equal. As in rule (a), ΔN_{bb} is maximized but with $\Delta N_{bb} \geqslant 0$, and the cells are scanned for movement as in rule (a).

Rule (iii)

This system is the same rule as rule (ii) except that the hexagonal cells are able to move to any of 36 neighboring locations (6 nearest, 12 next-nearest, and 18 next-to-next nearest neighboring locations).

Rule (iv)

This rule is the same as rule (iii) with the additional feature that w cells are also allowed to move so as to maximize ΔN_{ww}, but only if $\Delta N_{ww} \geqslant 0$.

Using the above rules, we simulate the cell movement for various fractions of b cells: $f_b = 0.1$, 0.2, 0.4, 0.6, and 0.8. For each of these fractions the simulation was carried out at least five times using each of the above rules. The results of the simulation are summarized in table 1. As in Goel et al. (1970), the average number of bb contacts per b cells is used as a measure of the extent to which b cells are clumped together. In column 1 is given the rule number, in column 2 we have given a brief description of each rule. For each fraction in the following columns we have given the average N_{bb}/N_b denoted by a, the standard deviation σ, and the average number of clusters formed, n. This latter figure is much more subject to chance variations than N_{bb}/N_b and thus it is not too useful for comparison of the various rules. For $f_b = 0.6$ and 0.8, we have not listed this number, because only very rarely do we get more than one cluster for any of the rules listed. In addition to the above results, several other descriptions of the movement were also calculated. Among these are the average displacement of a cell during the sorting, the average number of moves made by a cell, the maximum displacement of any cell, and the maximum number of moves made by any cell. These results will be given later in this section for some of the better rules.

We plotted average N_{bb}/N_b for the various rules as a function of f_b in figure 1 for square cells and in figure 2 for hexagonal cells. As shown in figure 1, we find that the graphs can be separated into three categories depending upon the number of the spaces to which moves are allowed. For example, if cells are restricted to exchanging only with their 8 nearest neighbor cells, only a small amount of improvement in aggregation is possible with any of the rules tried. We have arbitrarily taken rule 7 to represent this category, which corresponds to curve 2. The vertical bars around curve 2 denote the ranges of values for N_{bb}/N_b obtained by using all the rules tried in this study in which movement to only 8 neighboring places is allowed but that have one or more features different from rule 7. Curves 3 and 4 correspond to rules 11 and 20, respectively, in which movements to 20 and 36 neighboring places are allowed. The vertical bars around them have meanings similar to

Table 1.

f_b				0·1	
Rule no.	Rule description	a	σ	n
Ideal aggregate		1·675		
b	8 $>bb$ only	0·675		17·0
a	8 $\geqslant bb$	1·107	0·079	7·0
c	8 $\geqslant -1bb$	0·905	0·073	8·2
1	4 $\geqslant bb$ axes only	0·765	0·063	11·0
2	4 $\geqslant bb$ diagonal only	0·995	0·120	8·8
3	8 $\geqslant bb$ prefers same direction	1·137	0·031	6·5
4	8 $\geqslant bb$ random choice of cell to move	1·110	0·034	7·4
5	8 $\geqslant bb$ prefers direction, rescans moving cells	1·175	0·048	6·0
6	8 $\geqslant bb \geqslant ww$ w cells move also	1·180	0·048	6·0
7	8 $\neq <bb$ ΔN_{bb} not maximized but $\geqslant 0$	1·105	0·051	7·8
8	8 maximum bb increase; $\geqslant bb$ no scanning	1·242	0·077	5·3
9	8 $\geqslant bb$ number and size of blobs increases	1·130	0·070	7·0
10	20 $\geqslant bb$ prefers inner layer of moves	1·240	0·046	5·4
11	20 $\geqslant bb$ all moves equal	1·290	0·066	4·8
12	36 $\geqslant bb$ all moves equal	1·302	0·090	4·7
13	36 $\geqslant bb$ equal, prefers same direction	1·275	0·016	5·0
14	36 $\geqslant bb$ equal, prefers direction, rescans moving cells	1·380	0·123	3·8
15	36 $\geqslant bb$ prefers inner layer of moves	1·200	0·067	6·0
16	36 $\geqslant bb$ inner, prefers same direction	1·250	0·061	5·0
17	36 $\geqslant bb$ inner, prefers direction, rescans moving cells	1·355	0·037	4·0
18	36 $\neq <bb$ ΔN_{bb} not maximized but $\geqslant 0$	1·315	0·051	4·4
19	36 $\geqslant bb \geqslant ww$ inner, w cells move also	1·325	0·084	4·4
20	36 $\geqslant bb \geqslant ww$ equal, w cells move also	1·417	0·047	3·3
21	36 $\geqslant bb \geqslant ww$ equal, prefers direction, rescans moving cells	1·320	0·040	4·4
22	44 $\geqslant bb$ all moves equal, ΔN_{bb} not maximized	1·456	0·054	3·0
Ideal aggregate		2·450		1·0
(i)	6 moves $\geqslant bb$	1·720	0·156	5·6
(ii)	18 moves $\geqslant bb$, 2 layers all positions equal	1·910	0·013	4·0
(iii)	36 moves $\geqslant bb$, 3 layers all positions equal	1·965	0·070	3·6
(iv)	36 moves $\geqslant bb \geqslant ww$, positions equal, w cells	2·054	0·074	3·0

In column 2 the number given is the number of neighboring spaces to which the move is allowed and $\geqslant bb$ stands for $\Delta N_{bb} \geqslant 0$, $>bb$ for $\Delta N_{bb} > 0$, $\geqslant -1bb$ for $\Delta N_{bb} \geqslant -1$, $\neq <bb$ stands for ΔN_{bb} not maximized but $\geqslant 0$, etc. For detailed description of the rules, see text. a is the average number

Results of simulation

	0·2			0·4			0·6		0·8	
a	σ	*n*	*a*	σ	*n*	*a*	σ	*a*	σ
1·775			1·837			1·867		1·887	
0·912		21·0	1·287		11·0	1·467		1·672	
1·296	0·027	8·2	1·524	0·025	4·3	1·642	0·022	1·791	0·017
1·141	0·083	7·8	1·398	0·069	3·3	1·528	0·041	1·634	0·022
0·958	0·030	13·4	1·292	0·021	8·0	1·529	0·023	1·699	0·006
1·210	0·030	8·4	1·457	0·029	4·6	1·605	0·029	1·720	0·015
1·305	0·034	7·4	1·531	0·034	4·8	1·661	0·020	1·778	0·013
1·277	0·028	8·6	1·489	0·044	5·8	1·651	0·012	1·776	0·010
1·354	0·047	6·7	1·517	0·025	3·8	1·656	0·013	1·782	0·022
1·337	0·031	7·2	1·519	0·019	4·8	1·672	0·025	1·776	0·007
1·304	0·045	7·7	1·534	0·034	3·6	1·665	0·022	1·788	0·018
1·296	0·006	8·7	1·494	0·023	5·8	1·650	0·018	1·776	0·019
1·308	0·019	7·3	1·495	0·014	2·2	1·640	0·020	1·768	0·012
1·455	0·037	5·2	1·646	0·037	2·6	1·763	0·016	1·808	0·022
1·477	0·060	5·6	1·690	0·011	2·2	1·768	0·010	1·809	0·009
1·527	0·050	4·4	1·705	0·033	3·0	1·805	0·010	1·832	0·016
1·467	0·067	5·8	1·722	0·017	2·6	1·803	0·011	1·839	0·008
1·492	0·048	5·2	1·712	0·023	2·6	1·806	0·025	1·826	0·012
1·480	0·038	5·2	1·706	0·017	2·2	1·788	0·011	1·834	0·008
1·465	0·056	5·6	1·729	0·038	2·2	1·808	0·016	1·837	0·011
1·480	0·067	5·6	1·710	0·038	2·4	1·786	0·018	1·828	0·004
1·512	0·057	4·6	1·724	0·021	1·8	1·802	0·017	1·834	0·013
1·537	0·029	4·0	1·704	0·016	3·0	1·798	0·020	1·845	0·012
1·529	0·073	4·4	1·738	0·026	2·3	1·798	0·019	1·849	0·015
1·530	0·072	4·6	1·753	0·055	2·4	1·806	0·020	1·828	0·007
1·543	0·040	3·8	1·767	0·033	1·3	1·817	0·029	1·843	0·009
2·575		1·0	2·725		1·0	2·771		2·794	
1·971	0·048	6·7	2·330	0·064	3·4	2·556	0·033	2·676	0·022
2·096	0·071	5·3	2·372	0·056	4·8	2·591	0·029	2·693	0·018
2·194	0·065	4·3	2·522	0·048	2·8	2·681	0·019	2·746	0·022
2·208	0·035	3·7	2·605	0·026	2·2	2·760	0·006	2·751	0·011

N_{bb}/N_b, σ is the standard deviation, and *n* is the average number of clusters. For $f_b = 0·6$ and ·8, *n* is very close to 1 for all the rules. With the exception of rules (i), (ii), (iii) and (iv) (for which ·xagonal cells are used), square cells are used in the simulation.

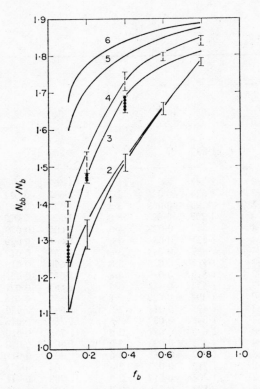

Fig. 1. N_{bb}/N_b in a 20 × 20 *square* array vs. the fraction of b cells f_b. Curves 1, 2, 3 and 4 correspond to rules (b), 7, 11 and 20 (see text). Rules (b) and 7 allow movement to eight neighboring spaces, 11 to 20 neighboring spaces, and 20 to 36 neighboring spaces. The vertical bars around curve 2 denote the ranges of values obtained by using the rules in which movement to only eight neighboring spaces is allowed, but which have one or more features different from rule 7. The vertical bars around curves 3 and 4 have similar meanings. The curve 5 is plotted using the best results we obtained by using any of the rules. Curve 6 represents the maximum possible values of N_{bb}/N_b for a given f_b. The curve 6 is plotted by using the relation $N_{bb} = 2[N_b - N_b^{\frac{1}{2}}]$ where the brackets signify truncation to an integer.

those around curve 2. From curves 2–4, we conclude that the greater the freedom given to a cell to move, the better are the chances of getting maximum possible value for N_{bb}/N_b (corresponding to a perfect aggregate). This view is further supported by the result that if the cell moves with maximum N_{bb} within the entire 20 spaces (rule 11), there is better sorting than if it prefers moves to the nearest 8 spaces (rule 10).

By applying statistical tests for the difference of means to the results of table 1, the following conclusions can be drawn for moves to 8 neighboring places.

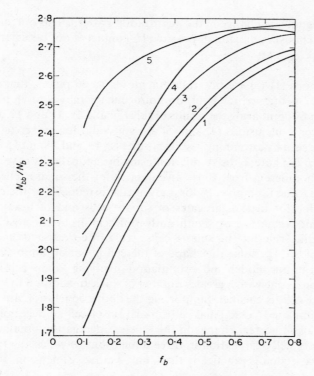

Fig. 2. N_{bb}/N_b in a 20×20 *hexagonal* array vs. the fraction of b cells f_b. Curves 1, 2, 3 and 4 correspond to rules (i), (ii), (iii) and (iv), respectively. Curve 5 represents the maximum possible value of N_{bb}/N_b for a given f_b.

(1) Directing a cell to prefer to continue moving in the same direction as the previous move (rule 3) results in a small improvement for the lower f_b values over rule (a), in which cells move randomly to the most favorable of the eight spaces.
(2) Allowing the cells that successfully complete a move to move again in the scan (rule 5) gives improved aggregation for the lower f_b values. However, it fails to improve aggregation for larger f_b values.
(3) Completely random choice of the cell to be moved (rule 4) as opposed to scanning results in poorer aggregation. However, this result may arise from a slowing down of aggregation rather than a change in the final pattern achieved. For monetary reasons, the program had to be cut off after a length of simulation comparable to that in other rules.
(4) Allowing both b and w cells actively to move (rule 6) also improves the aggregation for lower f_b. For higher f_b, the effect is negligible.
(5) If the move that a cell would make randomly, as long as $\Delta N_{bb} \geqslant 0$, is chosen (rule 7) rather than the move with maximum ΔN_{bb}, there is

no significant change in the aggregation pattern. This outcome implies that choosing maximum possible *bb* contact is not necessary for each step.

Conclusions (1)–(3) are not valid for moves to 36 places, that is, the improvements mentioned in (1)–(3) do not occur for moves to 36 places. This can be seen by comparing the results of rules 14 and 16, 15 and 17, and 21 and 20. However, conclusions (4) and (5) are still valid for moves to 36 places. This can be seen by examining results of rules 18, 13, and 14 and 15, 19, and 20.

As shown in figure 2, the simulation using hexagonal cells gave results not significantly different from those given above for the square cells. In fact, if the N_{bb}/N_b value for a given f_b for a given rule is reduced by the ratio of the N_{bb}/N_b value for ideal aggregates of square cells and of hexagonal cells, the resulting value is not significantly different from the results for the corresponding rule for the square cells; [rule (i) corresponds to (a), (ii) to 12, and (iii) to 21]. Thus, the shape of the cell is not an important facet of sorting out in this model, and a simulation involving square cells is as good a simulation as one with a greater number of sides. In addition the simulation with square cells is cheaper. Supporting the above conclusion are the results (a) of a "simulated" octagonal cell simulation of cell aggregation in which there was again no real change in the type of aggregation obtained, and (b) simulation of cell aggregation in three dimensions in which the two-dimensional cross section is practically the same as in two dimensions (Goel et al., 1970).

In figure 3 we have given some final pattern to rules 14, 20, and 21 for the square cells and to rules (ii) and (iii) for the hexagonal cells. A paradoxical observation may be made. The patterns with hexagonal cells have much greater visual similarity to sections of experimental aggregates in the rounding of edges of the islands. We should point out that the patterns given in figure 3 are the better patterns for a particular rule and not the average patterns. The standard deviation given in table 1 provides a measure for the difference between the best and the average. We have done so because we believe that the published experimental results of sorting out tend to include mainly the authors' best photographs and do not show trapping of cells although it is observed to occur. Thus we think that our better results correspond to the published photos of experimental sorting out. The patterns in which we found some trapping of *w* cells correspond to the unpublished but observed photographs.

There are three more remarks we would like to make about our simulation. First, the number of clusters of *b* cells on the average is usually maximum when the fraction of *b* cells is 0.2. Whether it is so experimentally or not, we do not know. Second, as expected, the average number of moves that a cell makes decreases with increase in fraction of *b* cells. The actual numbers are

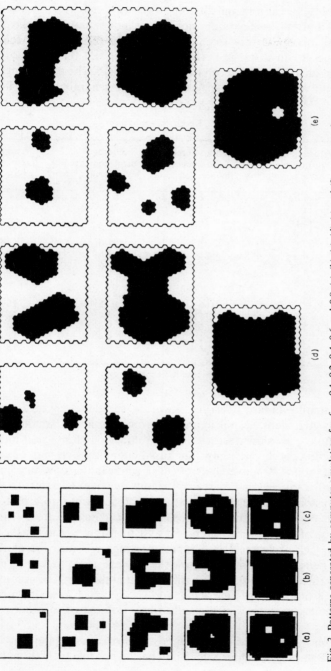

Fig. 3. Patterns generated by computer simulation for $f_b = 0.1, 0.2, 0.4, 0.6$ and 0.8. (a), (b), (c), (d), (e) and (f) correspond to rules 14, 20, 21, (iii) and (iv), respectively.

given in table 2 for rules 18 and 20 (rule 20 allows *w* cells also to move), where we have also given the maximum number of moves any cell makes and the average and maximum displacement made by a cell. By displacement we mean the Cartesian distance between the initial and the final positions if we take the side of the square cell to be of unit length. The maximum number of moves made by any cell varies greatly; the other numbers given in table 2

Table 2. Results of simulation

f_b	Average number of moves per *b* cell	Maximum number of moves by any cell	Average displacement of a cell	Maximum displacement of any cell
Rule 18				
0·1	15·0	31	5·6	15·0
0·2	11·0	27	5·1	13·0
0·4	7·0	25	4·6	13·2
0·6	4·3	21	3·5	12·0
0·8	2·15	24	2·0	11·3
Rule 20				
0·1	15·8	66	13·5	18·0
0·2	8·9	75	9·8	14·0
0·4	7·3	66	8·6	13·6
0·6	5·4	59	5·3	14·6
0·8	3·4	106	2·8	14·3

See text for explanation.

have very small standard deviation. Last, the simulated aggregation occurs in basically two steps. Rapid aggregation of cells into a small number of clumps followed by slow decrease in the number of clumps and increase in the compactness of these clumps. The rapid clumping step may be complete in one scan (about 80 movements of cells), whereas the slow step continues for about 50 scans. Beyond 50 scans, the improvement is extremely slow.

In conclusion, we have shown that cells will sort out into a pattern very close to the absolutely stable pattern as long as (1) the cells are able to "feel" the presence of cells several layers out and then move, and (2) the driving force is the lack of decrease in *E* function, which is a measure of adhesion among the cells. We do not know how the cells are going to feel the presence of cells, say three layers out. One possibility is that this might be accomplished by means of filopodia, pseudopodia, and so on, which have been reported to be present upon many of the cells when isolated or in monolayers on surfaces.

The simulation reported in this paper was carried out at the University of Rochester Computing Center, which is in part supported by National Science

Foundation Grant GJ-828. The authors are thankful to Dr. R. Gordon, Dr. R. Rosen, and Dr. M. Yčas for several useful discussions and for critical reading of the manuscript. Research was supported in part by the National Institute of Health Training Grant to the Department of Biology, University of Rochester. This research will constitute part of the thesis to be submitted by one of the authors (A. G. L.) in partial fulfillment of the requirements for the degree of doctor of philosophy.

REFERENCES

Goel, N., Campbell, R. D., Gordon, R., Rosen, R., Martinez, H., and Yčas, M. (1970) *J. Theor. Biol.* 28: 423.

Goel, N. S. and Leith, A. G. (1970) *J. Theor. Biol.* 28: 469.

Steinberg, M. S. (1970) *J. Exp. Zool.* 173: 395.

Trinkaus, J. P. (1969) *Cells into Organs*. Englewood Cliffs, N.J.: Prentice Hall.

10. Cell Aggregation Kinetics

P. Antonelli, T. D. Rogers, and M. Willard*

   This qualitative study provides a theoretical model of two-component cell sorting based on an exchange form of Steinberg's differential adhesion theory. The model differs substantially from those proposed by Leigh and Goel, who make use of an extended zone effect, thereby encouraging central clumping of the internally segregating component. Computer simulation predictions agree well with experimental results obtained by Trinkaus and Lentz but do not generally yield the central positioning often reported in other experiments. Possible reasons for this situation are discussed. In particular, it is suggested that central clumping is due to nonrandomness of the initial (aggregated) distribution, whereas the Trinkaus–Lentz results are due to randomness of the initial (aggregated) distribution. In Appendix A possibly serious mathematical errors in some of the Goel–Leigh models are discussed, whereas Appendix B proves the mathematical intractability of Steinberg's original three-dimensional model of cell sorting. In particular, a mathematical error that had led to a prediction contradicting experimental results of Roth and Weston is analyzed. It is also demonstrated that it is generally impossible to predict which of two cell types is the more adhesive merely by considering the geometry of final configurations in cell-sorting experiments.

1. INTRODUCTION

Steinberg has formulated the differential adhesion hypothesis for cell-sorting events in embryogenesis (Steinberg, 1964). This theory differs in several respects from the chemotaxic theory (see, for example, Edelstein, 1971) and from the timing hypothesis of Curtis (1967). Specifically, the Steinberg theory claims that the phenomena of cell sorting are explained by (1) random movement of individual cells and (2) differential adhesion among cell surfaces. Both (1) and (2) are stipulated to take place in the thermodynamic setting implied by requiring the more adhesive cell contacts to replace those of lesser adhesiveness.

The main purpose of the present study is to attempt to provide a theoretical model of cell sorting in which the character of the geometric and thermodynamic aspects inherent in the Steinberg theory can be fully realized. The thermodynamic aspect of our approach is based on the *exchange principle*, which is a natural consequence of the differential adhesion hypothesis. It dictates the motility rule for our model once the geometry of the model has

* First published in *J. Theor. Biol.* (1973) 41 : 1–21.

been determined. The geometry itself derives from the Steinberg hypothesis, which gives rise to a geometrical optimization problem that is (uniquely) solved by applying the general theory of packing for convex bodies in the plane. The unique solution dictates hexagonal-shaped model cells and a hexagonal grid as the natural choice for the geometry of the model. Our exchange model predicts entirely the Trinkaus–Lentz results (1964) but in general we do not obtain the central clumping effect of Steinberg (1964, 1970).

As another consequence of our model, we answer a question raised by Trinkaus (1969) as to why larger internally segregating clumps of an aggregate do not exhibit motility. In fact, we demonstrate that this property is a prediction of Steinberg's theory. It is also shown that when isolated our model cells exhibit random motion. This behavior has been observed experimentally (Ambrose, 1961; Gail and Boone, 1970; Trinkaus and Lentz, 1964).

The steady states of our model are seen to exhibit an interesting duality principle. For example, a 30% *B*-cell final state configuration is qualitatively similar to a 70% *B*-cell final state in that islands of *B* cells in the 30% experiment resemble islands of *A* cells in the 70% experiment much as a black and white photograph resembles the same photograph with black and white interchanged. This duality points up the general impossibility of determining which cell type (of two) is the most adhesive·from the geometry of the final state alone.

A study of this theory by Goel et al. (1970) motivated us in the present paper. In Goel et al. (p. 424), it is concluded that the theory of Steinberg in its original form was inadequate, and in Goel and Leigh (1970, 1971) it was endeavored to repair the theory by investigating a large number of so-called motility rules for cells. Leigh and Goel (1971, p. 187) state "We have shown that cells will sort out into a pattern very close to the absolutely stable pattern as long as (i) the cells are able to 'feel' the presence of cells several 'layers' out, and then move; (ii) the driving force is the lack of decrease of *E*-function which is a measure of adhesion between the cells." Although we agree heartily with (ii), we feel that the use of the extended zone effect of (i) may not be truly representative of the biological situation despite the observed pseudopodia in cases where individual cells can be seen. This is possible because the central clumping characteristics of Goel's "absolutely stable patterns" may not be typical final states of an aggregate.

This is certainly the case for the results of Trinkaus and Lentz cited above. Also, recent results of Elton and Tickle (1971) indicate that these may be experimental detection mistakes in sorting out phenomena so that central clumps might not actually be centrally located but rather more dispersed.* Yet our model is basically of exchange type so that the geometry of the initial distribution of aggregated cells determines to a large extent the final

* This research was supported by NRC Grant 7667 and NRC Grant 5210. We thank C. A. Tickle for explaining this fact to us.

arrangement (see figures in this paper). Thus, central clumps might result experimentally only because the initial aggregated distributions are not random, whereas the Trinkhaus–Lentz results may be due to the randomness of the initial aggregated distribution. The fact that Steinberg does not get centrally clumped islands for low percentages of the internally segregating component may be an indication of this effect.

In 1969 Trinkaus stated that differential adhesion appeared to be the most suitable hypothesis in the light of current experimental evidence (pp. 115–116). However, a then recent experiment of Roth and Weston (1967) gave strong evidence contradicting the Steinberg theory. Trinkaus states, "It appears therefore that the differential adhesion hypothesis has not withstood its first direct test ... (however) I suggest that the hypothesis not be abandoned yet, mainly because it explains so much."

There are several reasons why Roth and Weston (1967) do not contradict the differential adhesion theory (Steinberg, 1970). However, one of them is purely logical. The fly in the ointment is, quite simply, an erroneous mathematical argument (Steinberg, 1964, p. 348) from which the contradiction derives. In Appendix B we give the analysis of this error and its correction. As it turn out, the complete solution must be deferred because sphere-packing problems of a most difficult nature are inherent in Steinberg's approach. The insolvability of the simplest of these, at present, demonstrates the mathematical intractability of the three-dimensional Steinberg model.

A detailed description of our model is left to section 2. In section 3 we compare the model with others of Goel et al. (1970) and of Leigh and Goel (1971). The predictions of our model are described in sections 4 and 5. Appendix A discusses briefly a mathematical inconsistency in Goel et al.

2. Design of the Exchange Model

It is a familiar fact that the Euclidean plane can be regularly tessellated only by equilateral triangles, squares, and regular hexagons. One of these tessellations may be used for a computer study of cell-sorting phenomena. Goel (Goel et al., 1970) selected squares, whereas we shall select hexagons. The goal in either case is to design a two-dimensional model of the Steinberg theory that is amenable to simulation studies. The original model cells of Steinberg are three dimensional, and as is shown in precise fashion in Appendix B, such models are not tractable from a mathematical standpoint. We now justify our choice of hexagons.

Let us begin by supposing we have a certain fixed large number T of real aggregating cells of more or less the same size distributed in a certain region of space. For mathematical purposes we assume further that all cells have volumes close to a fixed positive number V and that they are approximately convex (i.e., any two points may be joined by a straight line entirely *in* the cell). Consider now the two-dimensional analog of this situation. Thus, cells are more or less convex planar disks, T in number, all of whose areas are close to a fixed positive number A. All models will be confined to a fixed region of the plane. If no cell contacts another, surely the sum of the perimeters for all the cells is a maximum. However, cells *do* contact one another. Indeed, because the Steinberg theory requires explicitly that weaker cell contacts be replaced by stronger ones, it is implicit that all cells would prefer contacts with other cells over contacts with medium.

Thus we come to the optimization problem: what is the resultant configuration of the aggregate if the total perimeter (i.e., sum of all cell perimeters) is minimized? A theorem from the geometry of planar convex bodies presented by Fejes Tòth (1964, p. 174) states that convex planar cells in maximal contact shall be circular in shape, of fixed radius and shall be arranged in a hexagonal beehive pattern. The theorem's hypothesis requires that each cell be convex and equal in area to any other. Actually, our model cells are not strictly convex and of equal area so that we are unable to apply the Fejes Tòth theorem with complete accuracy. However, it is reasonable to think that the configuration that solves the optimization problem for more or less convex cells of approximately equal area will be more or less of the hexagonal beehive pattern. Note that the optimizing cells are circular disks and not hexagons. It is the pattern of arrangement of these disks that is hexagonal. For purposes of modeling we are led via the Steinberg theory to hexagonal model cells in a beehive grid.

In this paragraph we follow the notation of Goel et al. (1970) and confine our analysis to two cell types, one of which may be medium. It is not difficult to generalize our procedure to more than two types of cells. The two cell types that compose the grid are randomly distributed. Of the two cell types, one will be denoted A and the other B. For a given cell, there are three types of contacts possible: AA, BB, and AB. To the edges of type AB, assign a non-negative number λ_{AB} representing force of adhesion. In a similar manner, define λ_{AA} and λ_{BB}. Note that we assume all edges of a given type have the same λ value. We define

$$F = \lambda_{AA}N_{AA} + \lambda_{AB}N_{AB} + \lambda_{BB}N_{BB'},$$

where N represents the number of edges of the indicated types. Steinberg's theory requires F to be maximized over the entire aggregate of T cells, the optimization being dependent upon the motility rules used.

Following Goel et al., we suppose throughout this discussion that λ_{BB} is greater than both λ_{AB} and λ_{AA} and that the number of B cells is less than the number of A cells. Thus, B cells will form the internally segrating component of the aggregate. In two-cell models, cells not internally segrating constitute the external component of the cell aggregate. For what follows, we define the term *total surface adhesion of a given cell* in an aggregate to be the sum of the λs for each of the six edges of a given cell.

Now the kinematics of our model follows from the requirement that AB adhesions be exchanged for BB adhesions, which have greater adhesiveness. Presumably, this exchange should occur in some kind of random way. In order to analyze the problem, we will assume tentatively that no cell is allowed to move to decrease its total surface adhesion. The following notation will be used. Each cell occupies a position in the hexagonal grid, and a typical position is denoted by an integer pair (i, j). We must distinguish between a cell and the position that this cell occupies. An A cell in the (i, j) position will be named a_{ij}. If this cell moves to some position (k, l), it will then be named a_{kl}. Likewise, B cells are denoted b_{ij}.

By the (i, j) *rosette* we mean the array of six cell positions surrounding the (i, j) position (see fig. 1). A_{ij}, the adhesive force of the (i, j) rosette, is defined to be the sum of the λs for the cell occupying the (i, j) position in the rosette. Thus A_{ij} is simply the total surface adhesion on whichever cell occupies the (i, j) position. By the (i, j) *rosette system* we mean the double layer of 18 cell positions surrounding the (i, j) position (see fig. 1).

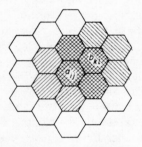

Fig. 1. The (i, j) rosette system.

Let us suppose as in figure 1 that an A cell occupies the (i, j) position and a B cell occupies one of the six adjacent positions, say (k, l), in the (i, j) rosette. Note that the (i, j) rosette and the adjacent (k, l) rosette overlap. We further suppose that cells satisfy the λ inequalities $0 < \lambda_{AA} < \lambda_{BB}$.

Let us suppose that the (i, j) rosette contains more B cells than the overlapping (k, l) rosette. Now, because we are tentatively assuming that no cell moves to decrease its total surface adhesion, the interchange $(i, j) \leftrightarrow (k, l)$ is allowable on the grounds that the total surface adhesion on b_{kl} increases, whereas it is not allowable on the grounds that the adhesion on a_{ik} decreases. Therefore, we now realize that not both cells in an interchange can increase their total surface adhesions. The next best thing would be to allow an interchange if the decrease in the total surface adhesion of one cell is small compared to the increase for the other cell. This would be the case, for example, if

λ_{AB} were much less than λ_{BB} and if $\lambda_{AA} = 0$. This relationship is the heart of the exchange principle, which we now formulate without any special conditions on the λs. Define

$$\Delta A_{lm} = A_{lm}^{\text{after}} - A_{lm}^{\text{before}}, \tag{1}$$

where A_{lm}^{after} is the adhesive force of the (l, m) rosette with a new cell in the (l, m) position as a result of a tentative exchange of the cell in the (l, m) position with one of the adjacent cells in the (l, m) rosette. We call an interchange $(i, j) \leftrightarrow (k, l)$ an allowable move if

$$\Delta A_{ij} + \Delta A_{kl} \geqslant 0. \tag{2}$$

This is the algebraic formulation of the exchange principle for our model. Statement (2) is mathematically equivalent to defining an allowable move as an interchange for which the increase in the total surface adhesion of one of the exchanging cells is greater than or equal to the decrease for the other cell.

It should be stressed that *not every interchange is an allowable move* even though it is always true that interchanging two adjacent A and B cells increases A_{ij} [A cell in (i, j), B cell in (k, l)] and decreases A_{kl}. For example, if the (k, l) rosette consists entirely of B cells and the overlapping (i, j) rosette has fewer than six B cells, then $(i, j) \leftrightarrow (k, l)$ is not allowable. [Such a move is not generally disallowed in the models of Goel et al. (1970) and Goel and Leigh (1970, 1971).]

Note that included among allowable moves are interchanges for which $\Delta A_{ij} + \Delta A_{kl} = 0$. Such moves, called 0 moves, play a fundamental role in cell aggregation phenomena in that they simulate random movement of cells. More will be said of 0 moves in later sections.

An allowable move is not necessarily a move that is actually made. Rather, there are six possible candidates in a given rosette for allowable moves. It may even occur that none of the six possibilities is allowable. In an actual computer run this possibility becomes more obvious as the system approaches stability. Our motility rule requires that of the allowable moves for a given rosette, one with maximal $\Delta A_{kj} + \Delta A_{kl}$ be made. The maximal one is selected at random in the cases where there is more than one such move.

3. Other Models

In Leigh and Goel (1971) 26 different motility rules (i.e., models) for cell sorting are studied. Three of these actually use a 20×20 hexagonal grid, whereas the remainder use a square grid of the same size. It is stated that results obtained for their models using hexagonal grids were much like those obtained using square grids. This may be true in our model to a certain extent too, but we have not checked it. In any case, none of the 26 models uses the exchange principle as far as we can tell.

As we have previously stated, the models of Leigh and Goel (1971) make use of extended zones of influence. Such models are not unlike chemotaxic models in their extended zone effect and would be expected to yield central clumping if the influence zone taken is large enough. Models of this type can certainly not be expected to predict the experimental results of Trinkaus and Lentz (1964) in which typical stable configurations exhibit no central clumping at all. This zone effect is minimal in the model we propose in that the only candidates for exchange are adjacent cells in overlapping rosettes.

Our model, on the other hand, indicates that central positioning is not a typical final state of the aggregate, and the results of Trinkaus and Lentz cited above corroborate this. The Trinkaus–Lentz experiment does not make use of a gyratory water bath shaker, whereas Steinberg recommends its use in cell-sorting experiments. Such a device produces a sustained vortex in the aggregating medium. The problem is that this motion may well induce central clumping of cells mechanically, as a function of their density independent of cellular adhesiveness. In order to check this point we recommend that a shaker device be used that better simulates random motion in space (as opposed to vortex motion) and that the results be compared with the gyratory shaker results. (This test and others have been carried out in the thorough analysis of Elton and Tickle (1971), which the reader should consult for more details).

Another interesting point is that models that do not employ some form of the exchange principle will necessarily possess the undesirable anticlumping trait of pulling internally segregating cells out of large clumps in order to form small nearby clumps. (See section 2 for an example of nonallowable interchanges.) In such situations the overall clumping effect is substantially reduced.

We are also able to formulate an answer to a question raised by Trinkaus (1969, p. 121) as to why large clumps of internally segregating cells do not show appreciable movement in experimental aggregates. In fact, as we shall see, such low motility for larger clumps is actually a mathematical prediction of the Steinberg theory. This is likely to be true for other models too.

As was stated earlier, most of the models of Leigh and Goel (1971) employ a square grid. In order to measure the effect of the hexagonal geometry of our models it would be interesting to make a computer study with hexagons by squares. We predict that the results will compare favorably with ours generally but that this agreement will falter in that cell clumps in a stable configuration will not be so compact as in the hexagonal model. Or, in Steinberg's terminology, the clumps of hexagons in a stable configuration will exhibit lower free surface energy than clumps of squares. This is precisely the thermodynamic content of the Fejes Tòth theorem.

Before continuing with the description of our results, we should like to point out what seems to be an inconsistency in the mathematical analysis in Goel et al. (1970) that may affect the entire paper. We refer the reader to Appendix A for a short discussion of this subject.

4. THE COMPUTER SIMULATION AND SOME PREDICTIONS OF THE MODEL

In a global study such as we feel is required by the Steinberg theory, such statistical quantities as "number of clumps in a final configuration" or "number of cells per clump" are not entirely appropriate. In what follows, we introduce a metric on the set of model cell configurations. This distance function is a version of the well-known Hamming metric of coding theory. We use this correlation metric to study the stability of cell aggregates in time, and we also use it to define a measure of metabolic energy consumed by an aggregate in passing from one configuration to another. Essentially, our correlation metric provides a natural language in which the global properties we study are easily expressible.

A total of about 1300 cells were used in a 40×40 hexagonal grid that tessellated a region of the plane with a two-layer boundary of A-type cells. Following Goel et al. (1970), we make the assumptions $\lambda_{AA} = 0 = \lambda_{AB}$ and $\lambda_{BB} = 1$. It should be emphasized that the zero values of λ_{AB} and λ_{AA} are not used to indicate that the A cells and B cells have no affinity for each other, but rather as a measure of the relative affinity as compared to BB contacts. The scanner chooses model cells at random and allows local exchanges of cells as dictated by the programmed motility rule. We decided to try to express the idea that in equal time intervals no real cell has more of an opportunity to move than any other, on the average. This property is called simultaneity and it means, for instance, that we disallow a model cell's having the opportunity of moving ten times while another has the chance to move only once during a small time interval. This does not imply that all cells move the same distance in a given run but only that they have equal opportunity to do so.

Because there are approximately 1300 cells in the grid, the scanner is allowed 1300 trials (not all result in cell movement), in which no internally segregating cell once moved is allowed to move again until 1300 trials have been made. This defines the time unit for our model. Thus, most of the cells in the grid have the opportunity to move once (and only once) per unit time. This particular measure of time occurs on our plots (see figures 2–13).

For comparison purposes we made nonsimultaneous runs and allowed one trial per unit time. Many other simulations of nonsimultaneity are possible. The resulting stable patterns are qualitatively similar to the simultaneous control-run stable configurations. However, it always happens that nonsimultaneous runs require less time to reach stable patterns than the simultaneous control runs.

We make constant use of the terms *stable configuration* or *stable pattern* to describe the final state of a run. In point of fact, final states are not stable at all but are rather steady states. Cell movement does not cease, but rather, the aggregate reaches a configuration after a time, about which it oscillates in a very regular fashion and with small amplitude [see figs. 2(a) and (b)]. The amplitude of oscillation seems to be independent of the relative number of *A* and *B* cells in the aggregate, a rather remarkable phenomenon. The moves made in the steady state are predominantly 0-moves. From time to time a nonzero move is made. If 0-moves are disallowed after the aggregate reaches equilibrium, no moves at all occur and oscillation ceases. In this sense, 0-moves set-up nonzero moves in the final state, but these nonzero moves are a very small minority of all occurring moves and are never enough to jar the configuration out of its steady state.

We first came across the importance of 0-moves in cell-sorting phenomena when they were routinely disallowed during a run in order to check the result against a control run allowing 0-moves. The final configurations reach stability in that very few moves are occurring per unit time in the final aggregate. There are a large number of single cells (as opposed to clumps) scattered throughout the stable configuration, and some fusion of larger cell clumps occurs. The overall geometry is seen to be similar to that of the initial

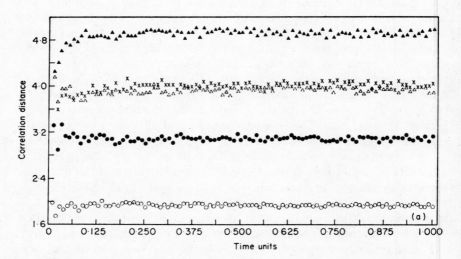

Fig. 2(a). Correlation distance from initial distribution vs. time is plotted exhibiting the time course of sorting out for the various $\bar{B}$ cell percentages. The correlation distance of a configuration from the initial state is determined at one-time-unit intervals. The steady state is reached in approximately 10 to 15 units of time for all cases. The duality principle (see text) is exhibited by comparing the 30% plot to the 70% plot. Percentage of *B* cells: ×, 70; ▲, 50; △, 30; ●, 20; ○, 10.

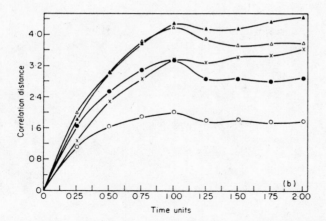

Fig. 2(b). Sorting out for the same computer runs as in figure 2(a) except that (b) displays only the first two time units of a run. The correlation distance is taken at quarter unit intervals thereby providing an interpolation of the plots of figure 2(a). Percentages of *B* cells: ×, 70; ▲, 50; △, 30; ●, 20; ○, 10.

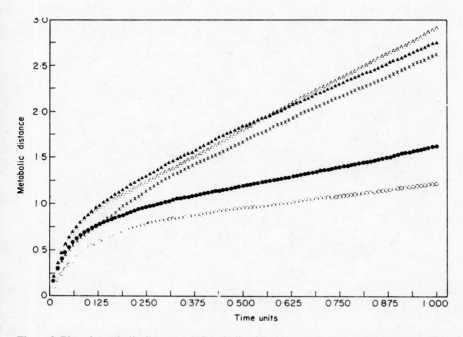

Figure 3. Plot of metabolic distance vs. time, indicating that metabolic energy (see text) is used up continually during the time course of sorting out and even in the steady state. The rate of energy use in the steady state is seen to be approximately constant as the slopes of the plots become constant. Percentages of *B* cells: ×, 70; ▲, 50; △, 30; ●, 20; ○, 10.

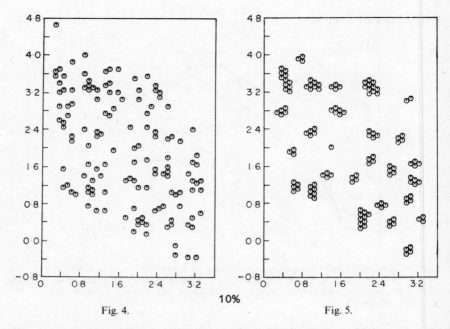

10%

Fig. 4.   Fig. 5.

Figures 4 through 13 exhibit typical initial and final state distributions of hexagonal model cells in the indicated *B* cell percentages. A position in the parallelogram-shaped grid is co-ordinated by the indicated axes. Only the positions of *B* cells, which are the most adhesive, are shown. A position not filled by a *B* cell is filled by an *A* cell. There are approximately 1300 positions on the grid. Figures 9 and 13 illustrate the duality principle (see text).

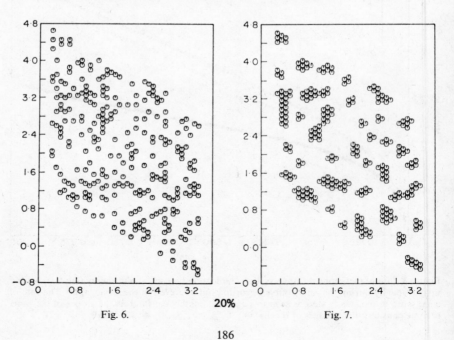

20%

Fig. 6.   Fig. 7.

186

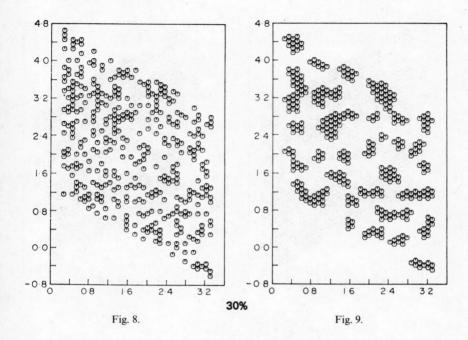

30%

Fig. 8.   Fig. 9.

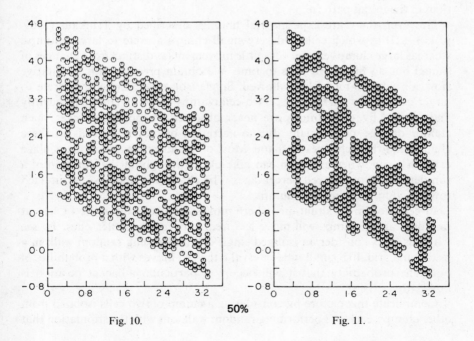

50%

Fig. 10.   Fig. 11.

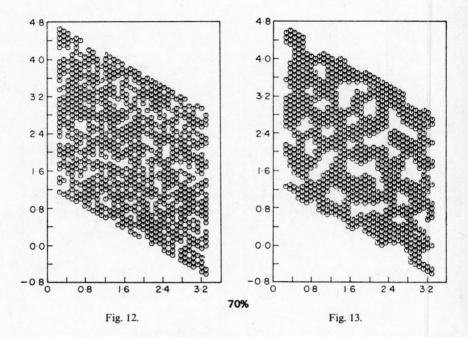

70%

Fig. 12.   Fig. 13.

pattern except for this fusion. Also, final *B*-cell clumps are more compact than in the initial pattern.

However, there are experimental findings described by Trinkaus (1969, p. 119–121) in which cells and very small clumps accrete to larger clumps, whereas large clumps exhibit very little movement (as distinct from changes of shape) but do fuse from time to time. We obtain precisely this qualitative situation when 0-moves are allowed. Single isolated model cells accrete to larger cell masses via 0-moves. Two-cell clumps move much less frequently and four or five-cell clumps are never observed to move. However, such clumps do fuse occasionally due to accretion of individual cells to these clumps and their changes in shape. Most of the accretive action takes place early in a run, whereas fusion can take place throughout the duration of a run depending on the cell percentages. The changes of shape seem directed toward compactifying cell clumps.

In cell movement situations where moves are occurring more or less at random, larger clumps will move less frequently than smaller ones. To see why this is so, consider an isolated single cell undergoing random walk in a hexagonal grid. In a small time interval t, the cell moves with a probability of one. The probability that it will occupy a particular adjacent position is one-sixth.

Complicate this picture by considering a clump of two cells isolated from other clumps, each cell performing random walk but with the stipulation that any movement of the clump must preserve its integrity. One easily calculates, with the use of figure 1, that the clump will move in time interval t with probability approximately one-third. One can make similar arguments for larger clumps. The general result is that the larger the clump the less likely it is to move.

The above discussion shows that the rarity of movement among the larger clumps follows essentially from the motility hypothesis of the Steinberg theory. The hypothesis asserts that single cells move. The model predicts that they move at random when isolated. This behavior has been observed in the laboratory (Ambrose, 1961; Gail and Boone, 1970).

To continue, we generally allow 0-moves to occur whenever they come up during a given run. One can argue that such moves should be allowed to occur only at random. To check this point, we performed several runs allowing 0-moves with probability $\frac{1}{2}$. The results are very similar to those for a control run in which 0-moves are performed with probability one. The only real difference is that the time required to reach stability is longer for the 50% 0-move runs.

Other rules are possible. For example, one could perhaps argue that all 0-moves that come up ought to be allowed at random toward the beginning of a run and that toward the end none should be permitted (or at least, be permitted with lower probability). In any case, there will be very little qualitative difference in the final states.

A typical output configuration has been chosen for our figures for approximately 10%, 20%, 30%, 50%, and 70% B cells. Recall that we employ always $\lambda_{AA} = 0 = \lambda_{AB}$ and $\lambda_{BB} = 1$, so that B cells will constitute the internally segregating component when the percentage of Bs is less than 50%. Initially, we found that a 30% B-cell run is like the 70% B-cell run in that in the final state, islands of B cells in the 30% run correspond in a strong qualitative way to the islands of A cells in the 70% run (see figs. 9 and 13). There is, however, the minor difference that the *dual A* islands were not quite so large, on the average, as the corresponding B islands. Evidently, this result is due to our use, following Goel et al. (1970), of a double layer of A cells for the boundary of our grid, so that a 70% B-cell run has a bit fewer than 30% A cells that can move. (The use of the double layer is merely a technical device to facilitate computer simulation.) This approximate duality principle explains why we have only considered percentages of 50% or less for B cells in our displayed figures.

We also studied an unusual type of stability that allows small perturbations of cell proportions during a run. A modification of the basic program is used in which a small percentage of B cells is replaced by A cells at random and at each unit of time. The general qualitative aspects of the final configurations resemble those of the unperturbed control runs. Final islands of B cells in the unperturbed control run are eroded or eaten away in the perturbed final configuration. By and large, the erosion is noticeable only near the boundaries of B-cell clumps or along thin bands of B cells connecting two larger B-cell islands.

When percentages of B cells are slightly increased instead of decreased, similar results are obtained. Thus, this type of perturbation study indicates that small variations in cell proportions in an aggregate undergoing cell-sorting do not appreciably alter the qualitative aspects of the final steady state.

In Goel et al. (1970) the notion of a "trapped cell" is introduced. Such a trapped cell would, for example, be an A cell in the interior of a B-cell clump in a stable configuration. We have found that in our model, trapped A-cell clumps do not occur when the B-cell percentage is less than 50%. For percentages greater than 50%, the previously described duality occurs and there are always trapped clumps of A cells.

To check more on this phenomenon we randomly scattered several A cells into a clump of internally segregated B cells and ran the basic computer program. Invariably, the cells wander out of the B-cell clump into the surrounding sea of A cells. However, if somewhat larger numbers of A cells are scattered through a B clump, trapped clumps of A cells do result. This effect is simply a miniature illustration of the duality principle.

Our usual condition on the λ's is $\lambda_{AB} = 0 = \lambda_{AA}$. However, the duality principle does hold in any case where λ_{AA} is substantially less than λ_{BB} (so that $\lambda_{AA} \approx 0$, relative to λ_{BB}). In fact, the principle holds, regardless of the value of λ_{AB}. This is so because interchanging A cells and B cells in a given pattern (excluding the boundary) will not appreciably affect the number of AB edges in the pattern. The boundary does, of course, produce a small change in the number of AB edges precisely because its double layer of A cells is not changed to a double layer of B cells in the exchange process. The principle of duality, approximate though it may be, is a prediction of our model of Steinberg's theory and might therefore be sought in the laboratory.

An important consequence of the duality principle is the impossibility of deciding which of two cell types in an aggregate has stronger adhesiveness from merely looking at the final stable configuration (with the A-cell boundary removed). It would seem this should be true of real aggregates as well. Certainly, this duality principle does not violate Steinberg's theory of cell sorting. For, if B cells have been determined to have greater adhesivity for one another than for A cells or than A cells for one another, the theory would predict that A cells will engulf B cells, provided of course that there are enough A cells (fewer than 50% A cells would in general be insufficient). The duality principle shows that the converse of this implication is false. That is, in general, it is impossible to conclude from the final configuration that B cells are more adhesive than A cells. Thus, the duality principle suggests that adhesivity of real cells cannot be determined generally by examining which cells engulf which in cell sorting. Other means must be employed.

5. Interpretation of the Plots

Suppose we are given two $m \times n$ binary matrices M and N, each with the same width (number of ls). Then, a finite number of transpositions (exchange of neighboring entries) will transform M into N. Let $d_m(M, N)$ denote the minimum number of transpositions necessary to convert M into N. It is easily checked that d_m is a distance function on the set of all $n \times m$ binary matrices of constant width. The relation to aggregation kinetics is simply this: cell movement consumes metabolic energy (Steinberg, 1964, p. 343) and the more movement, the more energy consumed. It is therefore of interest to compute the minimum energy required for a given aggregation to pass from one state to another and to compare this energy, in turn, to that actually consumed as a measure of efficiency of the kinetics. We are unfortunately not able to determine d_m. However, we introduce a measure of the energy actually consumed by an aggregate, which we call the metabolic energy. This energy, which we define in terms of the generalized Hamming metric, can be defined in other types of models. This metabolic energy may be close to d_m in models using an exchange principle.

The necessary metric is constructed as follows. Given two $m \times n$ binary matrices M, N, add them entrywise modulo 2, then compute the weight of the resultant binary matrix. One verifies that this procedure defines a metric d_c on the set of $m \times n$ binary matrices. The distance $d_c(M, N)$ between M and N simply counts the number of entries in which M and N differ. We call d_c the correlation metric.

As was stated, there is no clear relation between d_m and d_c. However, the (actual) metabolic energy consumed in passing from I_0 to I_i is conveniently related to d_c by the definition

$$M(i) = \tfrac{1}{2} \sum_{j=1}^{i} d_c(I_{j-1}, I_j),$$

where I_0 and I_i are the initial configuration and that at time unit i, each considered as a binary matrix. The A cells are represented by 0s and the B cells by 1s.

The graphs of $M(i)$ versus time i reveal that (1) rate of energy consumption becomes constant for large time, (2) these constant slope values are close for close cell percentages, and (3) dual distributions have approximately the same slope (see fig. 3). By comparing a plot point on these graphs with the corresponding binary matrix (i.e., grid), we are able to conclude that all model cell interactions after a sufficient length of time are due predominantly to 0-moves. This relationship applies to all runs.

The graphs of $d_c(I_0, I_i)$ versus time all exhibit, for the various cell proportions, a large initial increase [see figs. 2(a) and (b)] reflecting the large number of allowable moves in the beginning of a run. The amplitude of this initial jump varies directly with the proportion of B cells in an aggregate until the 50 % point is reached, at which time the duality principle predicts that it should decrease as the proportion of *B* cells increases. Note that the 70 % curve and the 30 % curve in figure 2(a) have comparable initial jump amplitudes, the discrepancy being attributable mainly to boundary effects. It should be noted that the correlation curves for different initial configurations are roughly congruent, reflecting qualitative similarity of cell sorting.

After the initial jump the curve settles into oscillating very regularly within a horizontal band of stability. As was previously mentioned, this oscillatory motion is due mainly to 0-moves. Also note that the metabolic curves become straight at about the same time the correlation curves enter their bands of stability.

Generally, then, we find that the model predicts relatively rapid formation of small internally segregating clumps, which then slowly enlarge by accretion for a time, with perhaps some fusion of large clumps, to finally settle down in a steady state in which there are few changes. The changes that do occur are minor changes in shape of the clumps and are mediated almost entirely by 0-moves. The large-scale geometry of the stable state can be seen in the large-scale geometry of the initial configuration in all cases. During the first phases of a run, isolated single cells exhibit random walk, whereas larger clumps exhibit essentially none but do exhibit fusion throughout most of period before stability. Cell islands change shape quite a bit but this process too settles down as the stable state is approached. It is significant that central clumping does not occur in our model unless the initial configuration is itself more or less central. Therefore, the central clumping results observed by Steinberg in the laboratory would not seem to be caused only by differential adhesion and cell motility but perhaps by the nonrandomness of the initial aggregated distribution as stated in the introduction.

We are indebted to many people for helpful criticism and discussion of the results and ideas of this paper. We would especially like to thank J. G. Timourian and S. Zalik of the University of Alberta, J. P. Trinkaus and C. A. Tickle of Yale University, and M. S. Steinberg of Princeton University. We are also indebted to E. Butz of the University of Alberta, who told us of the Goel papers and who encouraged us in the initial stages of this work. We would also like to thank the Computer Center at the University of Alberta for use of their facilities.

REFERENCES

Ambrose, E. J. (1961) *Exp. Cell Res.* (Suppl. 8): 154.
Curtis, A. S. (1967) *The Cell Surface: Its Molecular Role in Morphogenesis.* London: Lagos Press.
Edelstein, B. (1971) *J. Theor. Biol.* 30: 515.
Elton, R. A. and Tickle, C. A. (1971) *J. Embryol. Exp. Morph.* 26: 135.
Fejes Tòth, G. (1964) *Regular Figures.* New York: Pergamon Press.

Gail, M. H. and Boone, C. W. (1970) *Biophys. J.* 10: 980.
Goel, N. S., Campbell, R. D., Gordon, R., Rosen, R., Martinez, H., and Yčas, M. (1970) *J. Theor. Biol.* 28: 423.
Goel, N. S. and Leigh, A. G. (1970) *J. Theor. Biol.* 28: 469.
Leigh, A. G. and Goel, N. S. (1971) *J. Theor. Biol.* 33: 171.
Roth, S. A. and Weston, J. A. (1967) *Proc. Nat. Acad. Sci. U.S.* 58: 974.
Steinberg, M. S. (1964) *The Problem of Adhesive Selectivity in Cellular Interactions in Cellular Membranes in Development*, ed. M. Locke. 22nd Symposium of The Society for the Study of Growth and Development.
Steinberg, M. S. (1970) *J. Exp. Zool.* 173: 395.
Trinkaus, J. P. (1969) *Cells into Organs.* New Jersey: Prentice-Hall.
Trinkaus, J. P. and Lentz, J. P. (1964) *Develop. Biol.* 9: 115.

APPENDIX A

In a distribution of two types of square cells, let N_a and N_b denote the number of cells of type a and b and let N_{aa}, N_{bb}, and N_{ab} denote the number of edges of the indicated types. On p. 430 of Goel et al. (1970) it is claimed that

$$2N_{bb} + N_{ab} = 4N_b \tag{A1}$$

and

$$2N_{aa} + N_{ab} = 4N_a. \tag{A2}$$

These equations are called the conservation equations by Goel, who goes on to say, "In writing the conservation relations, we ... consider the boundary to be made up entirely of the predominant cell type." The boundary is further stipulated to be fixed, that is, not movable. It should be noted that equation (A1) above holds *only* when cells of type a make up the boundary. This means that equation (A1) should be used when a cells are predominant. Equation (A2), on the other hand, holds *only* when b cells make up the boundary, when b cells are in the majority. Counterexamples to show that equation (A1) does not hold when b-cell boundaries are employed are easily constructed, as are those to show that equation (A2) does not hold when an a-cell boundary is used.

It would then seem that simultaneous use of equations (A1) and (A2) would be inconsistent and lead, perhaps, to contradictions. However, equations (9a), (9b), and (9c) in Goel et al. are the basic equations for the mathematical analysis of two-cell patterns and are derived using equations (A1) and (A2) simultaneously. The analysis of three-cell pattern (ibid., pp. 432–33) seems to suffer from the same defect.

One can perhaps argue that boundary effects become insignificant in large aggregates so that results based on simultaneous use of equations (A1) and (A2) will be approximately true. This philosophy seems particularly reasonable for qualitative studies. For example, in a 100×100 square grid (with a surrounding boundary layer of cells of either type) 4% of the cells are adjacent to the boundary. However, Goel uses only 20 × 20 grids. So 20 % of the cells are adjacent to the boundary. This is certainly too large a percentage to be insignificant and so simultaneous use of equations (A1) and (A2) may lead to false conclusions. In light of the above discussion we therefore feel that considerable caution should be exercised by those interested in the results of Goel et al. (1970).

APPENDIX B

As was mentioned in the introduction, the experimental results of Roth and Weston (1967) are in direct contradiction to one prediction of Steinberg's theory. We will now show that this prediction follows from erroneous mathematical arguments and that, therefore, it is not actually a prediction at all.

Let us suppose, with Steinberg, that we are given an aggregate consisting only of cells of types a and b. It is stated that "a bond is established between contacting units (i.e., cells) at every point where a pair of adhesive sites are opposed" (Steinberg, 1964, p. 348). On p. 347 it is also stated that "sites of mutual attraction are distributed stochastically in the adhesive surfaces." We thus arrive at a picture of a typical model cell as spherical in shape with certain small randomly situated areas on its surface reserved as adhesive sites.

Denoting the frequency of adhesive sites per unit area on the cell surfaces of a- and b-type cells as f_a and f_b, Steinberg derives the work equations

$$W_a = k(f_a)^2, \qquad W_b = k(f_b)^2, \qquad W_{ab} = k(f_a)(f_b), \tag{A3}$$

where k is a constant of proportionality. Assuming that $W_a \geqslant W_b$, it follows trivially that

$$(W_a + W_b)/2 \geqslant W_{ab} \geqslant W_b. \tag{A4}$$

The Roth–Weston experiment showed, for a certain type of two-cell aggregate, that isotypic pairings between cells are preferred over heterotypic ones. Mathematically, this means

$$W_b > W_{ab}, \qquad W_a > W_b > 0, \tag{A5}$$

which yields the contradiction with equation (A4).

On p. 348 (Steinberg, 1964) it is stated that the probability of site apposition is given by $(f_a)^2$, $(f_b)^2$, and $(f_a)(f_b)$. Unfortunately, this statement is incorrect, and in addition, it is the key step in the derivation of equation (A3). For it is stated, "When very many sites are present, the number which are apposed will be directly proportional to the probability of apposition; and since the work of adhesion, W, is proportional to the number of sites which are apposed, we may introduce the proportionality constant, k, and write the equations (A3) above."

We select the quantity $(f_a)(f_b)$ to work with in the following discussion. We first note that this quantity is *not* the probability of *sb* site apposition in the aggregate at all. It is only the probability that a point of contact between two cells is an *ab* adhesive site. The correct expression for this probability is just the ratio of the number of *ab* sites to the total number of sites of all three types in the aggregate. The difficulty in determining this ratio is that it is strongly dependent on the local geometry of the aggregate, that is, on the manner in which model cells are packed together.

The analysis of the dependence of the probability of *ab* site apposition on the geometry proceeds as follows. We first suppose, for the sake of simplicity, that cell surfaces are nondeformable spheres, all of the same radius, and that the adhesive sites on the cells are circular patches fixed in number and diameter. Suppose further that each cell type has a characteristic number of sites per unit surface area.

Applying Steinberg's differential adhesion hypothesis we may suppose that the model cells are packed together in optimal fashion and that points of mutual contact are actually adhesive sites. In this fashion, a configuration in which the number of cells per unit volume is maximal is obtained. Clearly, the local geometry of this optimal configuration depends only on the relative strengths of the three types of adhesion. It is also evident that we cannot actually know the number of *ab* sites in the aggregate (and hence the probability of *ab* site apposition) without knowing the local geometry.

As an illustration of the difficulty of this question of the local geometry, consider the simplest case. This is the situation in which the three types of bonds have identical nonzero strengths. The resultant configuration is just the so-called densest packing of spheres (of the given radius). Unfortunately, the packing that is actually the densest is not at present known (Fejes Tòth, 1964). Therefore, we have no alternative but to conclude that the Steinberg model is mathematically intractable. Nevertheless, we feel that the present paper demonstrates how this problem can be circumvented.

11. A Rheological Mechanism Sufficient to Explain the Kinetics of Cell Sorting*

Richard Gordon, Narendra S. Goel,
Malcolm S. Steinberg, and Lawrence L. Wiseman

When two different vertebrate embryonic tissues are dissociated into individual cells, which are then recombined into mixed aggregates, the differing cells sort out within the aggregates to form a characteristic structure. The kinetics of cell sorting closely resembles the kinetics of breaking of an unstable emulsion of two immiscible liquids. We investigate the consequences of the postulate that cell rearrangement in such a system is driven by the tension at the interfaces between the two cell populations and resisted by tissue viscosities, the latter being a newly recognized parameter of cell sorting. Using preliminary experimental data on cell population interfacial tensions and on the time for fusion of two identical spherical aggregates, the viscous liquid model leads to estimates for tissue viscosities in the range of 0.4×10^6 to 0.7×10^8 poise. Also, using two other independent sets of data—one on the time for breaking of a roughly cylindrical cell aggregate into a few clusters and the other on the time for the rounding up of an approximately ellipsoidal tissue mass into a roughly spherical mass—tissue viscosities are again estimated to be in the range of 10^6 to 1.5×10^8 poise. In attempting to find a possible basis for such high effective viscosities, we propose that (1) tissue viscosities would most likely result from sliding friction between the cell membranes; (2) the cell membranes would have to possess protrusions of molecular or macromolecular dimensions; and (3) the ratio of the surface tension to the logarithm of viscosity should, if this model is correct, be approximately constant, independent of the tissue.

1. INTRODUCTION

When cells from two different vertebrate embryonic tissues are dissociated and recombined into a randomly mixed aggregate, they sort out, often approaching the configuration of a sphere of one type of tissue within a sphere of the other (Steinberg, 1963, 1964, 1970). The same configuration may be approached through a spreading or engulfing process after pieces of differing embryonic tissues are held in contact and fused (Townes and Holtfreter, 1955; Steinberg, 1962b, 1963, 1964, 1970). One of us (Steinberg,

---

* Note to the reader: This article is reprinted from the original [*J. Theor. Biol.* (1972). 37: 43–73], with the following improvements: (1) the references have been brought up to date; (2) illustrations cited from other papers are included (e.g. new fig. 1). The reference list is a nearly comprehensive bibliography on fluid systems whose change of shape is driven by surface tension and resisted by viscosity.

1962*b*, 1963, 1964, 1970) postulated that such behavioral traits of cell populations would follow directly from their possession of motility and of quantitative dIfferentials in adhesiveness. This connection has been formulated thermodynamically by Phillips (1969). The existence of the postulated adhesive differences among cell populations obtained from different tissues has been demonstrated directly by the sessile drop method (Phillips, 1969, and in preparation; Phillips and Steinberg, 1969, and in preparation) (new fig. 1).

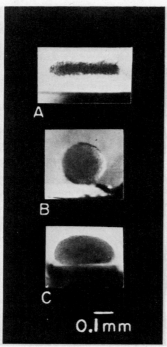

New Fig. 1. "Side-views of aggregates of four-day chick heart cells. (*a*) A flat aggregate freshly cut from a sheet of cells.... (*b*) A round aggregate formed by culturing a flat aggregate overnight on a gyratory shaker. (*c*) A flat aggregate following centrifugation at 4000 *g* for 24 hours. These aggregates each contain on the order of tens of thousands of cells." Figure 1 in Phillips and Steinberg (1969), with the authors' permission.

The mechanism of cell motility during cell sorting has not yet received the attention it deserves. In theoretical studies a lattice model was used that was originally devised to determine the stable configurations approached by model cell mixtures whose various types of interfaces were assigned a given set of adhesive strength values (Goel, Campbell, Gordon, Rosen, Martinez, and Yčas, 1970). However, when certain reasonable rules for cell motility were assumed in order to simulate the kinetics of cell sorting, the rearranging populations fell short of actually achieving the configuration of a sphere within a sphere (onion configuration, Goel et al., 1970), even though this was the most stable configuration. The cells became trapped in locally stable (metastable) configurations. With a different set of motility rules, there was a reasonable degree of cell sorting (Leith and Goel, 1971), but the clusters were not of minimum perimeter (cf. Vasiliev and Pyatetskii-Shapiro, 1971).

We thus learned that it is not sufficient to describe cells as motile and differentially adhesive to explain the dynamics of cell sorting. Their motility must be of a special kind. We were then faced with the problem of discovering a set of motility rules that not only give the observed final configurations at equilibrium but also produce kinetic behavior that allows these equilibrium configurations to be attained from a random mixture.

In lattice simulations, because of the nature of the models, a cell could make no movement less than a jump at least equal in length to its diameter. In this paper we will investigate the consequences of a continuous model in which cell rearrangement is driven by the tension at the interfaces between the two cell populations and is resisted by their tissue viscosities, the latter being a newly recognized parameter of cell sorting (Goel et al., 1970). In this model, all the cellular translocations are accompanied by decreases in interfacial free energy, but the cells move only tiny distances at a time. Combining preliminary experimental data on cell population interfacial free energies (Phillips and Steinberg, unpublished) and the time for fusion of two identical spherical aggregates, we obtain from this model estimates for tissue viscosities in the range from 0.4×10^6 to 0.7×10^8 poise. Also, using two other experimental observations, one on the breaking of a roughly cylindrical aggregate of cells into a few beads or clusters (Appendix B), and the other on the rounding up of an approximately ellipsoidal tissue mass into roughly spherical form (Appendix C), we derive from this model further independent estimates for tissue viscosities that compare very well with the estimate derived from the data on fusion. Using Eyring's absolute rate theory of viscosity (Glasstone, Laidler and Eyring, 1941), and assuming small protrusions of molecular or macromolecular dimensions on the membrane surfaces, we give a possible basis for such high viscosities in terms of the free energy of movement of protrusions past one another as cells slide past one another. Data are presented on the temperature dependence of the rate of cell sorting (Appendix D). The observed temperature dependence in the 30–37°C range is in good agreement with that calculated on the basis of the model.

A review of the kinetics of lattice models, including a crude lattice simulation of the continuous liquids model, is given in Appendix A. Appendices E and F discuss the interfacial tension between unlike tissues and relevant aspects of engulfing and spreading phenomena. We proceed directly with the description and consequence of the viscous liquid model of cell sorting.

2. Immiscible Viscous Liquid Analogy: A Continuous Model

In earlier papers (Steinberg, 1962*b*, 1963, 1964, 1970; Phillips, 1969), rearranging heterogeneous cell systems have been compared with immiscible liquid systems, and certain thermodynamic properties shared by the two kinds of systems have been identified as the sources of their closely similar behavior. It was pointed out that the kinetics of cell sorting is similar to the kinetics of breaking of an unstable emulsion of two immiscible liquids. We would like to extend and refine this comparison by adding reasons for postulating that such liquid-like cell systems may be highly viscous. We will investigate the consequences of assuming that cell sorting is driven by intercellular forces and resisted by the tissue viscosity. According to this model, the only intrinsic motility required of a cell is that it should be able to undergo continual small changes in its intercellular contacts. This would happen if the cells were deformable and their membranes were in continual slight motion. We will assume (without committing ourselves to this assumption) that this motion is passive Brownian motion, as in liquids (see section 5). This model requires all cellular translocations to be associated with decreases in the interfacial free energy of the cell population, but these translocations occur by only minute fractions of a cell's membrane area at a time, essentially continuously rather than by large jumps, as was the case in the lattice model.

It is difficult to determine whether a random mixture of cells of two types is behaving precisely like a mixture of two immiscible, highly viscous liquids because the hydrodynamic equations have not been solved for such a complicated geometry. Moreover, it is not at first clear just what measurements should be made to investigate this question. A number of qualitative features of such a system are apparent, and some of them may indeed be detected in Trinkaus's description of the consolidation of groups of retinal pigment cells within a translucent aggregate of heart cells (Trinkaus, 1969): "With a sufficient proportion of pigment cells, small clusters form at random internally within a few hours . . . . Clusters enlarge by accretion, the adhesive addition of individual cells and tiny clusters. As clusters enlarge, they often contact adjacent clusters and fuse. Clusters are constantly changing shape . . . . These form changes often cause adjacent clusters to contact each other and fuse. During the final phases of sorting out (after approximately 48 hours) . . . clusters now round up . . . . There is also a contraction of networks of interconnected clusters . . . . This causes some clusters to pull apart . . . . Clusters of pigment cells do not disaggregate, do not move." (See new fig. 2.)

3. Fusion or Sintering of Tissues and Estimation of Tissue Viscosity According to the Viscous Liquid Model

If the viscous liquid model were correct, tissue viscosities could be estimated by comparing the dynamics of cell populations with the dynamics of

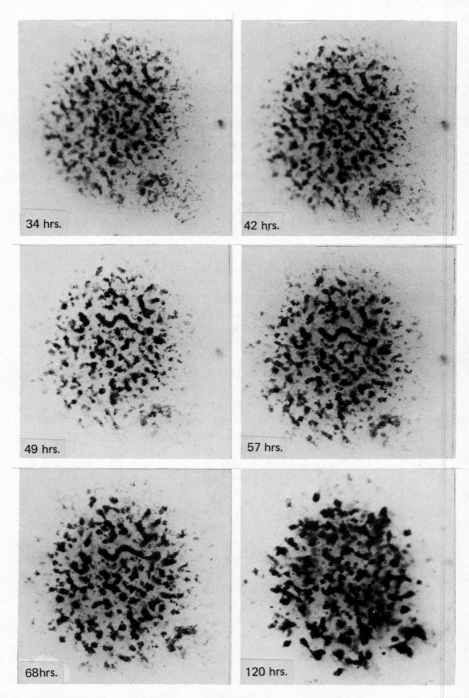

New Fig. 2. "Preparation number 5; proportion of pigment cells to heart cells is 1:4 ... 34 hours in culture. Elongate composite cluster is almost completely formed ... 5 days in culture. [120 hrs.] Several large clusters, which formed by fusion of smaller clusters, have separated into smaller clusters. Note especially several in the center of the aggregate ... Magnification: ×75." Reproduced from Trinkaus and Lentz (1964), figures 25–30, with permission.

viscous liquids placed in similar initial geometrical configurations. Various simple geometries are available for which the hydrodynamic equations have been solved for change of shape driven by interfacial tension and resisted by viscosity. We will discuss fusion (sintering) of two drops, the breakup of a cylinder into drops, the rounding up of an ellipsoid into a sphere, engulfing of one drop by another, and spreading of a drop on a surface (the latter four in appendices).

When two drops of a viscous liquid are placed in contact, they slowly fuse. The beginning of such fusion, or sintering, for isotropic, highly viscous liquids may be described by

$$x^2 = \frac{3a\sigma t}{2\pi\mu}, \tag{1}$$

where a is the radius of each of two identical, fusing spherical drops, t is the time, x is the radius of the circle of contact between the spheres, σ is the interfacial tension, and μ is the viscosity of the material constituting the drops (Frenkel, 1945).*

We took seven matched pairs of five-day embryonic chick heart re-aggregates (rounded up overnight after trypsin dissociation and reaggregation, Appendix D), and measured x at half-hour intervals after fusion. In figure 1 we have plotted $2\pi x^2/(3a)$ vs. the time t. The result is a straight line, as

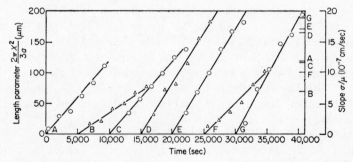

Fig. 1. Length parameter $2\pi x^2/(3a)$ (left scale) vs. the time t for the fusion of spheres of reaggregated chick cells. Each straight line is the least squares fit to a matched pair of average radii 148, 161, 159, 172, 159, 154, and 149 μ, respectively. The slopes, which measure σ/μ [equation (1)], are 11.8, 7.1, 11.7, 16.6, 17.2, 10.1, and 19.0 $\times 10^{-7}$ cm/sec (plotted on the right). The contact times, determined by extrapolation, are displaced by 5000-sec intervals for clarity. Matching of radii was within 10 μ.

predicted by equation (1), whose slope is to be interpreted as σ/μ. Least squares fits gave slopes from 7.1×10^{-7} to 19.0×10^{-7} cm/sec, with a mean of 13.4×10^{-7} cm/sec. Less extensive data on five-day liver reaggregates,

---

* Equation (1) clearly applies only to the beginning of fusion because x has a maximum $x_{max} = 2^{1/3}a \approx 1.26a$.

rounded-up liver fragments, and four-day forelimb bud cores gave 27.5 × 10^{-7}, 7.2 × 10^{-7}, and 4.4 × 10^{-7} cm/sec, respectively. Thus, in these preliminary experiments, σ/μ was found to range from 4.4 to 27.5 × 10^{-7} cm/sec. (The variability for identical tissues is not yet understood.)

Preliminary data from equilibrium sessile drop experiments (Phillips, 1969; Phillips and Steinberg, 1969) indicate that σ is in the range 1 to 30 dyne/cm (Phillips, personal communication). (Note that these are static measurements at shape equilibrium and are independent of viscosity.) Thus, μ is calculated from the above estimates for σ/μ to be in the range 0.4 × 10^6 to 0.7 × 10^8 poise. This is an exceedingly high viscosity, comparable to that of pitch at around 20°C (Hatschek, 1928). [See Lontz (1964) for a viscosity spectrum.]

Two other hydrodynamic-like phenomena have been similarly analyzed. In Appendix B we consider the breaking of a roughly cylindrical aggregate of cells into a few clusters. This is illustrated in a set of photomicrographs by Trinkaus and Lentz (1964, figs. 29 and 30) (see our new fig. 2), which show a rough cylinder of cells almost completely separating into three drops during a period of 52 hr. In Appendix C we consider the rounding up of an ellipsoidal tissue into a sphere which has been observed by Phillips and Steinberg (1969), who found that certain chick embryonic cell aggregates 0.008 × 0.04 × 0.04 cm^3 nearly rounded up overnight (new fig. 1) ($\simeq$ 16 hr, Phillips, personal communication). Drop formation gives a range of 5 × 10^6 to 1.5 × 10^8 poise, whereas the rounding up gives 1 × 10^6 to 3 × 10^7 poise for the tissue viscosity, according to the model. These data are in good agreement with the range estimated from the data on fusion.

Thus, we have demonstrated in three independent ways that tissue viscosities calculated in accordance with the viscous liquid model are remarkably high, with values in the range 0.4 × 10^6 to 1.5 × 10^8 poise. (Note that because the actual range of interfacial tensions may be less than the experimental estimate of 1 to 30 dyne/cm, the range of viscosities might likewise be more restricted.)

A liquid will change shape in response to any shearing stress, no matter how small. A viscous liquid takes some time to do this. Reaggregated tissues, placed in various configurations, slowly change their geometry. The final shapes suggest that they are behaving like liquids driven by interfacial tensions. When these interfacial tensions are opposed by another force of known magnitude, as by centrifugal force in the sessile drop experiments, their magnitude may be calculated from the final shapes.

The validity of the concept of tissue viscosity depends primarily on whether a tissue conforms to the above definition of a liquid. This property can be subjected to an experimental test. The only stresses that have been applied to reaggregated tissues are centrifugal forces and the tissues' own interfacial tensions. Other sources of stress, such as weights, could be imagined.

The cellular nature of the rearranging tissues, and the causes of the resistance to applied forces, are irrelevant to the calculation of tissue viscosities. But the viscosities we obtain are so uniquely high that it is reasonable to ask what cellular components could possibly be responsible for them.

The basic question we have to answer is why the cells rearrange so slowly. If they were moving actively and supplying a driving force over and above the interfacial tensions, the combined forces must then be meeting an even higher resistance (viscosity) than we have calculated. Thus we will assume, for purposes of the present argument, that the cells move passively.

The protoplasm inside various cells has a viscosity from 10^{-1} to 10^2 poise (Heilbrunn, 1958), which is many orders of magnitude too low to account for the calculated tissue viscosities. These measurements of protoplasmic viscosity presumably include the effects of microtubules and other structures within the cells. It is doubtful that cell membranes themselves could be sufficiently rigid to account for these tissue viscosities, especially in light of the results of Frye and Edidin (1970), which indicate that the membranes possess considerable fluidity. It is possible that a reaggregated tissue might be rigidly held together by a network of filaments, although such have not been observed.

We will next show that it is quite reasonable to suppose that such high viscosities reflect the difficulty of sliding the cell membranes past one another. If this should be the case, the tissue viscosity would strongly reflect the friction between the surfaces of the apposed cell membranes. There are three regimes for such friction (Akhmatov, 1966): (1) hydrodynamic friction, which is due to the viscosity of fluid between the surfaces; (2) juvenile friction, due to direct contact between surfaces; and (3) boundary friction, the intermediate case, which involves a layer of fluid whose properties are influenced by the surfaces.

Suppose that the sliding of the cell membranes involved hydrodynamic friction. One might regard the tissue as a concentrated emulsion of cells in the intercellular fluid. If the cells acted like an ordinary dispersed phase of spherical globules, high tissue viscosities would be predicted by empirical formulas that have been developed for such emulsions (Gillespie, 1963). On the other hand, cells in these tissues are flexible, so that it may be highly inaccurate to describe them as hydrodynamically interacting spheres. We are unaware of any viscosity measurements of a concentrated suspension, in a viscous medium, of flexible bags containing a low viscosity liquid. [Appropriately designed artificial cells (Chang, MacIntosh, and Mason, 1966) would be useful here.] It seems possible that the viscosity of such a suspension could be comparable to that of the viscous intercellular phase. If this were so, it appears that there is no plausible fluid to postulate for the intercellular medium that would both have a bulk viscosity of 10^6–10^8 poise and be water-immiscible. (If the medium were miscible with water, the solutes required to provide the high viscosity would draw water from the environs, and the viscosity would decrease.) Thus, it seems unlikely to us that the sliding friction is predominantly hydrodynamic, though rigorous proof remains to be given.

The fluid between the apposed cell membranes is likely to contain a large proportion of water whose structure will be influenced by the membranes. Peschel and Adlfinger (1970) have investigated the anomalous viscosity of water between hydroxylated fused silica surfaces. If, using their equation (5), we extrapolate their highest values to a membrane separation of 100 Å, the apparent viscosity of the water increases only fortyfold to 0.4 poise. Thus it seems unlikely that boundary friction could be responsible for high tissue viscosities. We therefore conclude that the high viscosity of sorting cell populations would most likely be due to juvenile friction involving direct contact between molecular constituents of the surfaces.* This we discuss in the next section.

4. A POSSIBLE EXPLANATION OF HIGH TISSUE VISCOSITIES

In this section we will provide a possible explanation for the high tissue viscosities (0.4×10^6–1.5×10^8 poise) we have calculated on the basis of the viscous liquid model. We shall use the classical Eyring theory of viscosity for ordinary liquids (Glasstone et al., 1941).

In Eyring's theory, viscosity of a liquid is calculated by using the theory of absolute reaction rates. This method is possible because the flow of a liquid is a rate process insofar as it takes place with a definite velocity under given conditions. The theory is developed in terms of the motion of one layer of molecules past another, which is assumed to involve the passage of a molecule from one equilibrium position to another such position in the same layer. This is possible if a suitable hole or site is available, which requires pushing back other molecules and, hence, expenditure of energy (new fig. 3). Thus, the jump of the moving molecule from one equilibrium position to the next is regarded as equivalent to the passage of the system over a potential energy barrier. By taking a symmetrical energy barrier, the following expression for the viscosity μ is derived [equations (22), (25), and (26), Chapter 9, Glasstone et al., 1941):

$$\mu = h\frac{N}{V}\,e^{\Delta F^{\ddagger}/RT}(\lambda_1/\lambda)^2, \tag{2}$$

* Parsegian and Gingell (1972, 1973) have attempted to model cell–cell interactions by using two thin lipid films, which represent the membranes (sometimes 'sugar' coated), and are surrounded by water on all sides. They calculate the van der Waals forces between the films using Lifshitz theory (Lifshitz, 1956). Their current model is continuous and uses the bulk properties of the phases. Thus if one imagines one membrane sliding relative to another, their calculations would yield hydrodynamic rather than juvenile friction and thus could not explain the high effective viscosities of the tissues. On the other hand, such van der Waals calculations could be repeated for bumpy membranes.

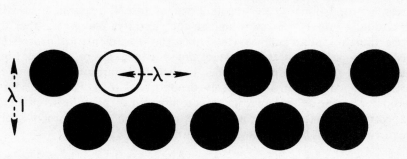

New Fig. 3. "Distances between molecules in a liquid; λ is the distance between two equilibrium positions for viscous flow." f indicates the direction of flow. Figure 118 in Glasstone et al. (1941), reproduced with permission of McGraw-Hill Book Company.

where h is Planck's constant $(=6.6 \times 10^{-27}$ erg sec), R is the gas constant $(=8.4 \times 10^{7}$ erg/mole), V is the molar volume of the moving molecules, N is Avogadro's number $(=6.2 \times 10^{23}$ molecules/mole) (so that V/N is the volume inhabited by a single molecule in the liquid state), $\Delta F^{\ddagger}$ is the standard free energy of activation per mole, T is the absolute temperature, λ_1 is the distance between centers of molecules in the two layers, and λ is the distance between two equilibrium positions in the direction of motion.

Now, in making the transition from considering the viscosity of moving liquids to considering the viscosity of moving cell populations, let us assume the existence on the apposed cell surfaces of tiny interdigitating protrusions that disengage and reengage, moving from one location to the next as two cells move past each other. For simplicity, let us take these protrusions to be spherical (radius s) and closely packed on the cell surfaces (cf. Appendix E). We shall also make the reasonable assumption that λ_1, the distance between the centers of these protrusions on apposed cells, approximately equals λ. Therefore equation (2) becomes

$$\Delta F^{\ddagger} = RT \ln \frac{\mu V}{hN}. \tag{3}$$

Note that because the dependence of $\Delta F^{\ddagger}$ on μ and V/N is logarithmic, $\Delta F^{\ddagger}$ is not very sensitive to either of them. Let us now estimate $\Delta F^{\ddagger}$ for this system. On the basis of experiments on cell aggregates, we have estimated above, through the viscous liquid model, a viscosity in the range of 0.4×10^{6}–1.5×10^{8} poise. If we choose a range 10–50 Å for s, the radius of a protrusion, and $T = 37°$ C $= 310°$ K, equation (3) yields a range of 14 to 23 kcal/mole for $\Delta F^{\ddagger}$. $(V/N = 4\pi s^3/3.)$

If the explanation given for the viscosity is correct, the tissue viscosity should decrease with an increase in T. More precisely, $\log \mu$ should be inversely proportional to the absolute temperature T. The time for sorting out should be directly proportional to the viscosity. We assume that this proportionality would be linear (this is true in the mathematical expressions used in this paper to describe fusion, time for breaking of a cylinder into clusters, and rounding up of a tissue), so that the logarithm of the time for sorting out would then be inversely proportional to T. (This, of course, would not be true if the change in temperature also changes the properties of the protrusions on the membrane to the extent that $\Delta F^{\ddagger}$ changes appreciably.) Thus, from equation (3), for two temperatures T_1 and T_2,

$$\log \frac{t(T_1)}{t(T_2)} = \log \frac{\mu_1}{\mu_2} = \frac{\Delta F^{\ddagger}}{R}\left(\frac{1}{T_1} - \frac{1}{T_2}\right) \tag{4}$$

or

$$\frac{t(T_1)}{t(T_2)} = \exp\left[\frac{\Delta F^{\ddagger}}{R}\left(\frac{1}{T_1} - \frac{1}{T_2}\right)\right], \tag{5}$$

where $t(T)$ is the sorting-out time at temperature T. From equation (5), for the two values of $\Delta F^{\ddagger}$, 14 kcal/mole and 23 kcal/mole, ratios of values of t for $T_1 = 30°C, 20°C,$ and $10°C$, and $T_2 = 37.5°C$ are as shown in table 1.

Table 1. Calculated values of the ratio of times for sorting out at different temperatures

$\Delta F^{\ddagger}$ kcal/mole	$t(30)/t(37.5)$	$t(20)/t(37.5)$	$t(10)/t(37.5)$
14	1.8	3.9	9.2
23	2.5	9.1	36.0

Sorting out of chick embryo four-day limb precartilage from five-day liver at these temperatures is described in Appendix D. From these experiments we conclude that $t(30°C)/t(37.5°C)$ for the combination studied, $\simeq 1.5$–2.0, is in agreement with the prediction given above. However, $t(20°C)/t(37.5°C)$ and $t(10°C)/t(37.5°C)$ are much larger than the predicted value. This situation does not in itself invalidate the viscous liquid model because $\Delta F^{\ddagger}$ may increase as temperature decreases and thus slow the sorting-out time.

More agreement between consequences of the viscous liquid model and experimental results is found if we carry the comparison of tissues with liquids a bit further. We will show that by knowing $\Delta F^{\ddagger}$ and applying the viscous liquid model, an estimate of surface tension can be made that is in accordance with the experimental value of 1–30 dyne/cm used in arriving at the viscosity. Let us first describe the procedure used for liquids (Glasstone et al., 1941).

Consider a liquid consisting of N molecules bound to each other by bonds with energy $NE/2$. $E/2$ is the energy required to vaporize a molecule, provided the other molecules adjust in such a way that no holes are left in the liquid. Because E is the energy required to vaporize a molecule when a hole is left, $E - E/2 = E/2$ is also the energy needed to make a hole of molecular size in the liquid without vaporizing a molecule. Therefore, the amount of energy required to make a hole in the liquid large enough for a molecule is the same as that needed to evaporate a molecule without making a hole. However, the latter quantity is equivalent to the energy of vaporization per mole ΔE_{vap}, where $\Delta E_{vap} = \Delta H_{vap} - RT$. ΔH_{vap} is the normal latent heat of vaporization, and RT is the correction for the external work done in vaporizing one mole of liquid, the vapor being assumed to behave as an ideal gas. For a molecule to take part in flow, a hole must be available. This hole is not necessarily the full size of a molecule but will be some fraction represented by the additional volume required by the activated state compared with the initial state. Therefore, the free energy of activation $\Delta F^{\ddagger}$ may be expected to be some fraction of the energy required for making the hole, that is, some fraction g of the energy $E/2$ required to vaporize a molecule without leaving a hole. For ordinary liquids g is about $\frac{1}{2}$ to $\frac{1}{4}$ (Glasstone et al., 1941). If we presume that this fraction g is a fundamental quantity, then knowing $\Delta F^{\ddagger}$, we can calculate the binding energy E from $\Delta F^{\ddagger} = gE/2$.

The surface tension is approximately half the free energy per unit area of cohesion, or $\sigma = E/4 = \Delta F^{\ddagger}/(2g)$ (Krupp, 1967). Therefore, for ordinary liquids, surface tension is about one or two times $\Delta F^{\ddagger}$ per unit area.

In our case, for two cells to move past each other, a protrusion has to move to the position of its neighboring protrusion. Again assuming the above-mentioned relationship between $\Delta F^{\ddagger}$ and E, $\Delta F^{\ddagger}$ per unit area will approximately be the free energy of cohesion per unit area and the surface tension will be half of it. For $\Delta F^{\ddagger} = 20$ kcal/mole and for an area of about 1.6×10^{-14} to 35×10^{-14} cm^2 per protrusion (area per protrusion for a hexagonal close packed arrangement $= 3\sqrt{3}s^2/4 = 1.55\,s^2$), corresponding to $s = 10$ and 50 Å, respectively, $\Delta F^{\ddagger}$ per unit area varies from about 90 erg/cm^2 to 4 erg/cm^2 (1 kcal $= 4.186 \times 10^{10}$ erg). Therefore, surface tension, according to this model, would be roughly 2–90 dyne/cm, which is consistent with the experimental values we have been using.

Using the same assumptions about the membrane protrusions and their interactions, it is possible to carry the theory yet another step and to predict interfacial tensions between different tissues from their individual surface tensions (Appendix E), provided the adhesion mechanisms are very similar for different cells. However, data are not yet available to quantitatively test this extension of the theory.

5. DISCUSSION

We have explored certain consequences of assuming a rheological mechanism of cell movement during sorting. According to the model investigated, the force responsible for sorting is the interfacial tension between two cell populations, and the force opposing the sorting is a frictional force. We have used the concept of tissue viscosity, introduced by Gordon (Goel et al., 1970), as a quantitative representation of this latter force. On the basis of this viscous liquid model, we have made three independent estimates of tissue viscosities. These estimates rely upon the experimental data on the rate of fusion of tissues, rounding up of nonspherical aggregates, and breaking of elongated multicellular strands into clusters—all of which are believed to occur due to the same force. All these estimates fall within the range of 0.4×10^6 to 1.5×10^8 poise. In an effort to find a possible basis for such a high effective viscosity, we concluded that it would most likely be due to sliding friction between apposing cell membrane surfaces, and that the cell membranes would have to possess interdigitating protrusions of molecular or macromolecular dimensions on their surfaces to experience such high frictional forces. It should be emphasized that equilibrium aspects of cell sorting could by themselves have afforded no basis for proposing the existence of irregularities on the cell membrane surfaces. This is so because, in principle, by choosing appropriate parameters and using standard theory of surface tension of liquids (Fowkes, 1967), one could account for the observed surface tension of cell populations even by assuming that the membranes have a smooth surface structure. Only the consideration of kinetic aspects of cell sorting has allowed us to differentiate between certain consequences of the presence of smooth versus bumpy cell surfaces. In light of the data considered here, the viscous liquid model seems to require the existence of bumpy cell surfaces. The required protrusions may already have been seen by electron microscopy (Spycher, 1970). If our rheological model has validity, then *ab initio* calculations of intracellular forces, modeling two neighboring cells as smooth parallel plates (Parsegian and Gingell, 1972, 1973), are not quantitatively applicable to the dynamics of cell sorting (see footnote on p. 204).

An interesting consequence of the rheological viewpoint is that the cells need not be motile in a classical amoeboid fashion (Steinberg, 1962a). Weston and Abercrombie (1967) observed, in fusions of labeled and unlabeled tissue fragments, that the cells seemed immobile despite the ability of the fragments to round up. They suggested that "cells within these fragments might establish adhesions by means of cytoplasmic projections which they continually extend and retract ... the translocation of cells might be associated with the tendency of more stable ('stronger') adhesions of these projections to increase their area at the expense of less stable ('weaker') ones. Displacement of the cells in these tissues may occur, therefore, only as a consequence of the process of maximizing adhesions as postulated by Steinberg." Trinkaus (1969) also suggested that "it may be necessary to substitute changes in cell shape, such as assumption of dendritic form and contraction of filopodia, for the motility component in Steinberg's scheme." The differential adhesion hypothesis does not prescribe the nature of the motility component, and it was early pointed out that the behavior of sorting units is "independent of the causes of their motility" (Steinberg, 1963, p. 403). However the motility rules can influence the kinetics of sorting. Indeed, the sources of the forces causing relative motion between cells might well vary from one real-life case to another (Steinberg and Wiseman, 1972). Nevertheless, it is sufficient (although not necessary) to model cell sorting as a passive process. The cells need only be able to change shape, but these shape changes themselves could be driven by interfacial tensions and require no metabolic energy. This behavior is in harmony with the previous statement by one of us (Steinberg, 1962b) that "energy is not expended for the rupture of a given intercellular adhesion if the break is induced by the force applied by another cell in the act of replacing the previous adhesion with a stronger one." The later statement (Steinberg, 1964) that "cell movement *per se* will consume metabolic energy under any circumstances: energy which must be expended in order to rupture existing attachments" was intended to refer only to active cell movements. Brownian motion due to thermal energy should in principle be sufficient to produce ruptures among existing attachments so that cell sorting could in principle be a passive process.

The observation that the drug cytochalasin B, an inhibitor of active cell motility, prevents cell sorting in cell mixtures of several kinds (Wiseman and Steinberg, 1971; Sanger and Holtzer, 1972; Maslow and Mayhew, 1972; Steinberg and Wiseman, 1972; Armstrong and Parenti, 1972) suggests that active cell movements are important in cell sorting. On the other hand, the observation that mixtures of neural retinal and pigmented retinal cells can sort out to an appreciable extent despite the presence of cytochalasin B (Armstrong and Parenti, 1972) suggests that the interfacial forces that evidently guide cell sorting are strong enough, in certain cases, to drive it as well. Thus it appears that the cell movements accompanying cell sorting result from a combination of active and passive processes ("cooperative" cell movement, Steinberg and Wiseman, 1972).

Of course, nothing in our presentation argues against the possibility that sorting out may in fact be accompanied by active cell motility (Garrod and Steinberg, 1973; Steinberg and Wiseman, 1972). There are three plausible modes for such motility: (1) *synergistic* motility, in which the cells actively enhance their motion in the same direction they would go if moved passively by interfacial tensions; (2) undirected, or *random* motion of the cell membranes additional to that caused by Brownian motion; (3) *antagonistic* motility, in which the cells actively oppose the physical forces generated by interfacial tensions. [Chemotactic motility of at least one kind has been eliminated by Steinberg (1962a).] Both (1) and (3) require an act of recognition of neighboring cells or of the direction of passive movement by a cell, and a coupling of this recognition to the mechanochemical system generating motility. These features go well beyond the minimal assumptions of the differential adhesion hypothesis.

Random motility, however, requires no cellular communication or transduction. Equation (2) is based on the assumption that thermal energy is the only energy available to move the protrusions past one another. An additional source of random energy ε may be described by adding ε to RT in equation (2), which would lower the viscosity and increase the rate of cell sorting.

In conclusion, we propose some additional experiments related to the viscous liquid model, some of which can provide tests of the model. These experiments are as follows.

(1) As pointed out above, one consequence of the immiscible liquid model is that active motion by the cells may not be necessary to produce cell sorting. We may add that according to the liquid model, random mixtures of wholly artificial cells (Chang, MacIntosh, and Mason, 1966), which can be made the same size as living cells, ought to sort out if they have appropriate surface properties (see also Chang, 1972).

(2) Because $\log \mu$ is proportional to the activation energy $\Delta F^{\ddagger}$ [equation (3)] and the binding energy E is proportional to σ, $\sigma/\log \mu$ should be approximately independent of tissue, if E is proportional to $\Delta F^{\ddagger}$, with constant of proportionality independent of tissue (which is approximately the case for true liquids as discussed above). A corollary of this statement is that if the surface tension of one tissue is larger than that of another, their viscosities should have the same ordering. Measurement of viscosities and surface tensions of a variety of tissues would be required to test this prediction of the model.

(3) In an initially random reaggregate of b and w cells, the b cells form a set of loosely interconnected networks (Curtis, 1967). What would be the average number of cells per network as a function of f_b, the fraction of b cells in the population? This problem is formally identical to the percolation bond problem (Hammersley and Handscomb, 1965; Gordon 1975, which asks how many interconnected, randomly distributed holes an artificial sponge must have to get the inside wet. The remarkable result is that there is a critical concentration of b cells below which there are many networks and above which there is essentially only one. If we now assume that the final rounded clumps are more or less directly related to these initial networks, it follows that for a given aggregate size and shape there will be a critical concentration f_b^* above which a single clump of b cells will usually be obtained. Although

f_b has been varied considerably (Steinberg, 1962a; Trinkaus and Lentz, 1964; Wiseman, Steinberg, and Phillips, 1972), no systematic attempt has been made to ascertain the existence or value of the critical f_b^*.

This critical concentration could be empirically determined by studying the final drop size in an artificial system consisting of two intimately mixed, immiscible, highly viscous liquids of equal density (which do not form a stable emulsion) as a function of the proportions of the two liquids. In a real cell-sorting system, the critical concentration should be the same as for the analog liquid experiment if cell motion is passive. Active motility would reduce the critical concentration. This effect seems manifest in the results of Armstrong and Parenti (1972), who observed that in neural retinal-pigmented retinal cell mixtures sorting out in the presence of the drug cytochalasin B, "in contrast to control aggregates which show single internal pigmented retinal masses, multiple retinal clumps are present at the completion of sorting."

(4) If the experiment on rounding up of ellipsoidal tissues described in Appendix C can be done accurately, then by studying the time course of rounding up and using equation (C2), the time dependence, if any, of σ/μ can be determined. If the cell surface properties and cell activity are not changing, σ/μ ought to be independent of time. Sintering can also yield the same information.

If the cells in a tissue are differentiating with respect to their adhesiveness, σ/μ should change (because from the argument presented above, $\sigma/\log \mu$ may not change). Thus, the time course of the rounding up of an ellipsoid to a sphere could also be used as a probe for the timing of adhesive changes in cells during embryogenesis.

(5) It should be possible to measure tissue viscosities independently of the interfacial tensions. This might be done, for instance, by measuring the velocity of a metal ball being slowly pulled magnetically through a layer of cells. The viscosity is given by Stokes's formula:

$$\mu = F/(6\pi r v), \tag{6}$$

where F is the applied force (gravitational plus magnetic), r the radius of the ball, and v its velocity. The thickness of the layer should be at least a few times greater than $2r$. (Alternatively, from such independently measured tissue viscosities, surface tensions could be calculated from fusion experiments and other methods.)

As a final remark, we conceive of a tissue hydrodynamics emerging, which will be distinct from the mainstream of hydrodynamics in that it will deal with liquids: (1) that are highly viscous; (2) that are sometimes visco-elastic; (3) in which the properties of the units may change; (4) whose units may be active suppliers of energy (Jacobson and Gordon, 1973); and (5) whose units may have anisotropic adhesive properties (Steinberg, 1964;

Goel and Leith, 1970). The main function of such a morphodynamics will be to derive the dynamic geometric consequences of the states and activities of embryonic cells (Jacobson and Gordon, 1975).

We would like to thank Robert Bender, Hsin-kang Chang, James F. Danielli, David Gingell, Karen Gordon, Gabor T. Herman, John Musgrave, Edward Paul, V. Adrian Parsegian, Herbert Phillips, Robert Rosen, and Martynas Yčas for enlightening discussions about cell sorting and for suggesting improvements in the manuscript.

This work was supported in part by the State University of New York through the Einstein Chair budget of Professor C. H. Waddington, NASA Grant NGR-33-015-016 to the Center for Theoretical Biology, NIH Grant 1-R01-HD-05136-01 to Robert Rosen, and NSF Grants GB5759X to one of the authors (M. S. S.) and GU-4040 to the Institute for Fundamental Studies. Revision was made at the Mathematical Research Branch, NIAMDD, National Institutes of Health, Bethesda, Md.

REFERENCES

Akhmatov, A. S. (1966) *Molecular Physics of Boundary Friction.* Jerusalem: Israel Program for Scientific Translations.

Armstrong, P. B. and Parenti, D. (1972) *J. Cell Biol.* 55: 542–53. Cell sorting in the presence of cytochalasin B.

Chang, T. M. S. (1972) *Artificial Cells.* Springfield, Illinois: C. C. Thomas.

Chang, T. M. S., MacIntosh, F. C., and Mason, S. G. (1966) *Can. J. Physiol. Pharmacol.* 44:115–28. Semipermeable aqueous microcapsules. I. Preparation and properties.

Cherry, B. W. and Holmes, C. M. (1969) *J. Colloid Interface Sci.* 29: 174–76. Kinetics of wetting of surfaces by polymers.

Curtis, A. S. G. (1967) *The Cell Surface: Its Molecular Role in Morphogenesis.* London and New York: Logos/Academic Press.

Dettre, R. H. and Johnson, R. E., Jr. (1970) *J. Adhesion* 2: 61–63. The spreading of molten polymers.

Fowkes, F. (1967) *Surfaces and Interfaces. I. Chemical and Physical Characteristics*, ed. John J. Burke, p. 197. Syracuse: Syracuse University Press.

Frenkel, J. (1945) *J. Phys. (USSR)* 9: 385–90. Viscous flow of crystalline bodies under the action of surface tension.

Frye, L. D. and Edidin, M. (1970) *J. Cell Sci.* 7: 319–35. The rapid intermixing of cell surface antigens after formation of mouse-human heterokaryons.

Garrod, D. R. and Steinberg, M. S. (1973) *Nature* 244:568–69. Tissue-specific sorting-out in two dimensions in relation to contact inhibition of overlapping.

Gillespie, T. (1963) The effect of concentration on the viscosity of suspensions and emulsions. In *Rheology of Emulsions*, ed. P. Sherman, pp. 115–24. New York: Pergamon Press.

Glasstone, S., Laidler, K. J., and Eyring, H. (1941) *The Theory of Rate Processes*, p. 480. New York: McGraw-Hill.

Goel, N., Campbell, R. D., Gordon, R., Rosen, R., Martinez, H., and Yčas, M. (1970) *J. Theor. Biol.* 28: 423–68. Self-sorting of isotropic cells. (Reprinted in this book, pp. 100–144.)

Goel, N. S. and Leith, A. G. (1970) *J. Theor. Biol.* 28: 469–82. Self-sorting of anisotropic cells. (Reprinted in this book, pp. 145–158.)

Good, R. J. and Elbing, E. (1970) *Ind. Eng. Chem.* 62: 54–78. Generalization of theory for estimation of interfacial energies.

Gopal, E. S. R. (1968) Principles of emulsion formation. In *Emulsion Science*, ed. P. Sherman. New York: Academic Press.

Gordon, R. (1975) Monte Carlo methods for cooperative Ising models. In *Cooperative Phenomena in Biology*, ed. G. Karreman. New York: Academic Press, in press.

Gordon, R. and Drum, R. W. (1970) *Proc. Nat. Acad. Sci. U.S.* 67: 338–44. A capillarity mechanism for diatom gliding locomotion.

Hammersley, J. M. and Handscomb, D. C. (1965) *Monte Carlo Methods*. London: Methuen.

Harvey, E. N. and Shapiro, H. (1941) *J. Cell Comp. Physiol.* 17: 135–44. The recovery period (relaxation) of marine eggs after deformation.

Hatshek, E. (1928) *The Viscosity of Liquids*, chap. 11. London: G. Bell and Sons.

Heilbrunn, L. V. (1958) *The Viscosity of Protoplasm, Protoplasmatologia. II., Cl.* Wein: Springer.

Isenberg, I. (1953) *Bull. Math. Biophys.* 15: 73–81. Cell division by swelling stresses.

Jacobson, A. G. and Gordon, R. (1975) Shape changes in the developing vertebrate nervous system analyzed experimentally, mathematically and by computer simulation (to be submitted to *Develop. Biol.*).

Krupp, H. (1967) *Advan. Colloid Interface Sci.* 1: 111–239. Particle adhesion, theory and experiment.

Leith, A. G. and Goel, N. S. (1971) *J. Theor. Biol.* 33: 171–88. Simulation of movement of cells during self-sorting. (Reprinted in this book, pp. 159–175.)

Levich, V. G. (1962) *Physicochemical Hydrodynamics*, chap. 5. Englewood Cliffs, N.J.: Prentice-Hall.

Lifshitz, E. M. (1956) *Sov. Phys. JETP* 2: 73–83. The theory of molecular attractive forces between solids.

Lontz, J. F. (1964) Sintering of polymer materials. In *Fundamental Phenomena in the Material Sciences*, vol. 1, ed. Bonis, L. J. and Hausner, H. H., pp. 25–47. New York: Plenum Press.

Marsland, D. (1956) *Int. Rev. Cytol.* 5: 199–227. Protoplasmic contractility in relation to gel structure: temperature-pressure experiments on cytokinesis and amoeboid movement.

Maslow, D. E. and Mayhew, E. (1972) *Science, N.Y.* 177: 281–82. Cytochalasin B prevents specific sorting of reaggregating embryonic cells.

Meister, B. J. and Scheele, G. F. (1967) *A.I.Ch.E.J.* 13: 682–88. Generalized solution of the Tomotika stability analysis for a cylindrical jet.

Miesse, C. C. (1955) *Ind. Eng. Chem.* 47(9): 1690–1701. Correlation of experimental data on the disintegration of liquid jets.

Parsegian, V. A. and Gingell, D. (1972) *J. Adhesion* 4: 283–306. Some features of physical forces between biological cell membranes.

Parsegian, V. A. and Gingell, D. (1973) A physical force model of biological membrane interaction. *A.C.S. Symposium on Adhesion, Washington, D.C., September, 1971*, ed. H. Lee. In *Recent Advances in Adhesion*, pp. 153–92. London: Gordon & Breach.

Peschel, G. and Adlfinger, K. H. (1970) *J. Colloid Interface Sci.* 34: 505–10. Viscosity anomalies in liquid surface zones. IV. The apparent viscosity of water in thin layers adjacent to hydroxylated fused silica surfaces.

Phillips, H. M. (1969) *Equilibrium Measurements of Embryonic Cell Adhesiveness; Physical Formulation and Testing of the Differential Adhesion Hypothesis*. Ph.D. thesis. The Johns Hopkins University, Baltimore, Maryland.

Phillips, H. M. and Steinberg, M. S. (1969) *Proc. Nat. Acad. Sci. U.S.* 64: 121–27. Equilibrium measurements of embryonic chick cell adhesiveness. I. Shape equilibrium in centrifugal fields.

Phillips, H. M., Steinberg, M. S., and Lipton, B. H. (1972) *Am. Zool.* 12: 699. Elasticoviscous morphogenetic behaviour of centrifuged embryonic chick cell aggregates.

Rashevsky, N. (1960) *Mathematical Biophysics* vol. 1, 3rd rev. ed., pp. 185, 217, 269. New York: Dover.

Rawles, M. E. (1963) *J. Embryol. Exp. Morph.* 11: 765–89. Tissue interactions in scale and feather development as studied in dermal-epidermal recombinations.

Rayleigh, Lord (1892) *Phil. Mag.* 34: 145–54. On the instability of a cylinder of viscous liquid under capillary force.

Sanger, J. W. and Holtzer, H. (1972) *Proc. Nat. Acad. Sci. U.S.* 69: 253–57. Cytochalasin B: effects on cell morphology, cell adhesion and mucopolysaccharide synthesis.

Spycher, M. A. (1970) *Z. Zellforsch. Mikroskop. Anat.* 111: 64–74. Intercellular adhesions, an electron microscope study on freeze-etched rat hepatocytes.

Steinberg, M. S. (1962a) *Proc. Nat. Acad. Sci. U.S.* 48: 1577–82. On the mechanism of tissue reconstruction by dissociated cells. I. Population kinetics, differential adhesiveness and the absence of directed migration.

Steinberg, M. S. (1962*b*) *Proc. Nat. Acad. Sci. U.S.* 48: 1769–76. On the mechanism of tissue reconstruction by dissociated cells. III. Free energy relations and the reorganization of fused, heteronomic tissue fragments.

Steinberg, M. S. (1962*c*) *Exp. Cell Res.* 28: 1–10. The role of temperature in the control of aggregation of dissociated embryonic cells.

Steinberg, M. S. (1963) *Science, N.Y.* 141: 401–08. Reconstruction of tissues by dissociated cells. (Reprinted in this book, pp. 82–99.)

Steinberg, M. S. (1964) The problem of adhesive selectivity in cellular interactions. In *Cellular Membranes in Development*, ed. M. Locke, pp. 321–66. New York: Academic Press.

Steinberg, M. S. (1970) *J. Exp. Zool.* 173: 395–434. Does differential adhesion govern self-assembly processes in histogenesis? Equilibrium configurations and the emergence of a hierarchy among populations of embryonic cells.

Steinberg, M. S. and Wiseman, L. L. (1972) *J. Cell Biol.* 55: 606–15. Do morphogenetic tissue rearrangements require active cell movements? The reversible inhibition of cell sorting and tissue spreading by cytochalasin B.

Szabo, G. (1955) *J. Pathol. Bacteriol.* 70: 545. A modification of the technique of "skin splitting" with trypsin.

Tomotika, S. (1935) *Proc. Roy. Soc. London A* 150: 322–37. On the instability of a cylindrical thread of a viscous liquid surrounded by another viscous liquid.

Torza, S. and Mason, S. G. (1969) *Science, N.Y.* 163 : 813–14. Coalescence to two immiscible liquid drops.

Torza, S. and Mason, S. G. (1970) *J. Colloid Interface Sci.* 33: 67–83. Three-phase interactions in shear and electrical fields.

Torza, S. and Mason, S. G. (1971) *Kolloid-Z.u.Z. Polymere* 246: 593–99. Effects of the line tension on 3-phase liquid interactions.

Townes, P. L. and Holtfreter, J. (1955) *J. Exp. Zool.* 128: 53–120. Directed movements and selective adhesion of embryonic amphibian cells.

Trinkaus, J. P. (1967) Morphogenetic cell movements. In *Major Problems in Developmental Biology*, pp. 125–76. New York: Academic Press.

Trinkaus, J. P. (1969) *Cells into Organs*, chap. 7. Englewood Cliffs, N.J.: Prentice-Hall.

Trinkaus, J. P. and Lentz, J. P. (1964) *Develop. Biol.* 9: 115–36. Direct observation of type-specific segregation in mixed cell aggregates.

van Oss, C. J. and Gillman, C. F. (1972) *J. Reticuloendothelial Soc.* 12:283–92. Phagocytosis as a surface phenomenon. I. Contact angles and phagocytosis of non-opsonized bacteria.

Vasiliev, A. B., Pyatetskii-Shapiro, and Radvoginn, Y. B. (1972) *Ontogenez (Akad. Nauk SSSR)* 2:356–62. Translated in this book, pp. 46–77.

Weston, J. A. and Abercrombie, M. (1967) *J. Exp. Zool.* 164: 317–24. Cell mobility in fused homo- and heteronomic tissue fragments.

Wiseman, L. L. and Steinberg, M. S. (1971) Abstr. 11th Ann. Mtg. Amer. Soc. Cell Biol., New Orleans, p. 328. The reversible inhibition of cell sorting and tissue spreading by cytochalasin B.

Wiseman, L. L. and Steinberg, M. S. (1973) *Exp. Cell Res.* 79:468–71. The movement of single cells within solid tissue masses.

Wiseman, L. L., Steinberg, M. S., and Phillips, H. M. (1972) *Develop. Biol.* 28: 498–517. Experimental modulation of intercellular cohesiveness: reversal of tissue assembly patterns.

Young, G. (1939) *Bull. Math. Biophys.* 1: 31–46. On the mechanics of viscous bodies and elongation in ellipsoidal cells.

APPENDIX A

Simulation of Cell Sorting Using Lattice Models

A population of cells is modeled in two dimensions by a space tessellated or subdivided into equal squares (a lattice). Each square can be occupied by a cell of any type. Each edge of contact between two cells is assigned an affinity λ, which is a measure of the strength of binding between the cell types. Thus, λ_{bb} and λ_{bw} are the affinities between cells of type b and b and between b and w, respectively. We also define an E function for a pattern of cells as the sum of the number of edges multiplied by their λ values. For a system with two types of cells (b and w),

$$E = \lambda_{bb}N_{bb} + \lambda_{bw}N_{bw} + \lambda_{ww}N_{ww}, \tag{A1}$$

where N_{bb} is the number of bb edges, and so forth. The equilibrium pattern has maximum E. It can be shown (Goel et al., 1970) that for isotropic cells (cells with λ independent of the directions of the neighbors) if

$$\lambda_{bb} > \tfrac{1}{2}(\lambda_{bb} + \lambda_{ww}) > \lambda_{bw} > \lambda_{ww}, \tag{A2}$$

then the equilibrium pattern is the one for which N_{bb} and N_{ww} are maximum and N_{bw} is minimum, that is, the pattern in which b cells form a single mass surrounded by w cells. For the purpose of studying motility we chose, for simplicity, $\lambda_{bb} = 1, \lambda_{ww} = \lambda_{bw} = 0$ so that from equation (A1) the equilibrium configuration is a square of b cells unpenetrated by w cells.

The rearrangement of randomly distributed cells of two types b and w was simulated using a FORTRAN computer program representing a set of plausible motility rules. Specifically, a certain fraction (f_b) of a 20×20 square grid was filled in random positions by b cells, and the rest of the grid was considered to be occupied by w cells. This arrangement resulted in each cell sharing edges with four̄ other cells except for those cells at the edge of the grid. Each cell was examined in turn and a decision made as to whether it should switch places with one of its up to eight neighboring w cells. The decision was based on the increase ΔN_{bb} in the number of bb pairs that would result. The w cell that would yield the maximum ΔN_{bb} was chosen. (If there were two or more such w cells, one of them was chosen at random.) However, ΔN_{bb} was required to exceed a certain minimum, $\Delta N_{bb}^{\min}$, for any switch to be made at all.

This last rule was adopted because when $\Delta N_{bb}^{\min} \leqslant 0$, an isolated b cell would be free to wander around in a mass of w cells. The extent to which such behavior occurs in living aggregates is under investigation (Wiseman and Steinberg, 1973). Published observations give little evidence for such cell wandering (Trinkaus and Lentz, 1964; Weston and Abercrombie, 1967). Using the above procedure, an initially random mixture of w and b cells never reached the equilibrium configuration (a square of b cells surrounded by w cells) with $\Delta N_{bb}^{\min} = 1, 0, -1, -2$, or -3 in any of the computer simulations (Goel et al., 1970); that is, not even when undirected cell wandering was permitted. The same conclusion was arrived at even for three-dimensional lattices in which each cell has 26 neighboring cells with which to switch.

When we allowed cells to move to slightly more distant locations than nearest neighbor locations (Leith and Goel, 1971), the local trapping decreased significantly and, in fact, if we allowed moves to next-nearest and next-to-next nearest locations, the final structure was all b cells clumped together (for $f_b \gtrsim 0.4$) but generally not in a square form. These simulations indicate that as the distance increases up to which a cell can "sense" the presence of its neighboring cells and then move, the self sorting improves. However, the stable equilibrium pattern of minimum perimeter is not generally obtained. This seems to be a general, inevitable property of lattice models with a finite range of interactions (cf. Vasiliev and Pyatetskii-Shapiro, 1971).

We have found another set of rules, simulating the transmission of forces over an indefinite distance that occurs in liquids, which gives clusters of minimum perimeter. As in the previous simulations with the lattice model (Goel et al., 1970; Leith and Goel, 1971), b and w cells are assumed to be on a square grid. However, a b cell (the internally segregating kind) is allowed to push into its clump and push another b cell out somewhere else on the edge of the same clump. This effect was accomplished through a FORTRAN computer program, which is available upon request.* This crudely simulates the continuous viscous liquid model of section 2 because, as in a liquid, local stresses are transmitted through the tissue. We have thus indefinitely increased the distance up to which a cell can sense the other cells and move in response.

The program has the following features. The b cells are initially placed at random with probability f_b (or in a compact rectangle to simulate breaking of a line of cells) on a planar square array. As before (Goel et al., 1970), a two-cell border of w cells is maintained. Each b cell is considered during consecutive scans but in random order within a scan. A b cell is not moved if it is isolated from other b cells or is completely surrounded. These rules are an attempt to reduce random migration, which we wish to avoid in the passive liquid model, and to prevent the formation of holes. If a b cell has two or three nearest neighbor b cells, it is moved only if such action would not isolate one of these neighbors. This rule attempts to reduce excessive breakup of clumps. Isolated pairs are not allowed to move, a rule that reduces the migration of small clumps. (The need for these special rules emphasizes the roughness with which this simulation approximates the liquid model. A full hydrodynamic numerical calculation would require an enormous investment in time and money.)

If the b cell is allowed to move, a search is made for an appropriate place for another b cell in the same clump to pop out. All w locations that are nearest neighbors of b cells in the clump are listed. The location that would give the b cell placed there the maximum number of b nearest neighbors is then determined. If there are two or more such locations, one is chosen at random. A b cell is placed, and the first b cell, imagined to push into the clump, is replaced by a w cell. The choice of a site with the maximum number of nearest neighbors for the b cell popping out guarantees that the clump will tend toward minimum perimeter.

* R. Gordon, Mathematical Res. Branch, NIAMDD, NIH, Bethesda, Md. 20014.

The topology of the clumps is not considered in calculating the location for a b cell to appear. Thus a hole within a clump would tend to fill up. To indicate when this artifact occurs, a b cell is forbidden to pop into a location with four nearest neighbor b cells.

The computing time for this program was excessive. More sophisticated pattern recognition techniques than those indicated above could speed such programs considerably. The program was run enough to find that essentially one clump formed when $f_b = 0.6$ on a 20×20 array, whereas four clumps appeared when $f_b = 0.5$, suggesting a critical value of $f_b \approx 0.6$ for a two-dimensional system. A different value, but the same phenomenon, should be anticipated for three dimensions. (The computer program is easily generalized to three dimensions, but in its current form the computing time would be prohibitive.)

Configurations for $f_b = 0.1$ to 0.5 after ten scans are shown in figure 2(a). The sequences of events for $f_b = 0.3$ and 0.6 are shown in figures 2(b) and (c). All the events described by Trinkaus in the quote in section 2 may be observed to happen in these simulations. (The migration of small clusters is an artifact of the rules.)

Figure 2(d) shows the breakup of a line of cells into separate "drops," a rough analog of the breakup of a cylinder of fluid, discussed below.

APPENDIX B

Breaking of a Cylindrical Aggregate into Clusters

In this appendix we will make an estimate of the tissue viscosity by comparing the breaking of a roughly cylindrical aggregate of cells into a few clusters with the breaking of a long cylinder of any fluid, suspended in another liquid, into a number of drops. This hydrodynamic instability phenomenon occurs under many conditions (Levich, 1962).

The case of a highly viscous cylinder of fluid in free space was analyzed by Rayleigh (1892). He concluded that "when viscosity is paramount, long threads do not tend to divide themselves into drops at mutual distances comparable with the diameter of the cylinder, but rather give way by attenuation at few and distant places." Tomotika (1935) extended the analysis to a viscous cylinder embedded in an immiscible viscous fluid. He found that such a cylinder should break up into equally spaced drops of a definite size, which is a function of the ratio of the viscosities only. Let μ_b be the viscosity of the cylinder and μ_w the viscosity of the surrounding medium. For instance, when $\mu_b/\mu_w = 1$, Tomotika finds that the spacing ought to be 5.58 times the diameter of the cylinder. A plot of the spacings versus μ_b/μ_w is given by Miesse (1955). The spacing has a minimum of 5.33 diameters at $\mu_b/\mu_w = 0.28$, and rises to 7.57 and 7.68 diameters for $\mu_b/\mu_w = 0.01$ and 10.0 (new fig. 4).

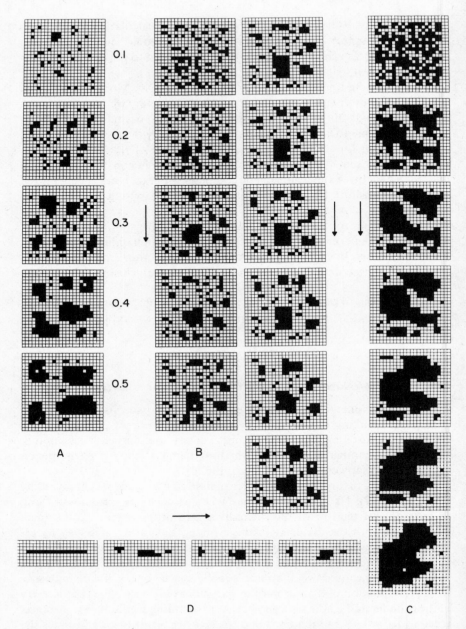

Fig. 2. Computer simulation of cell sorting by the viscous liquid model. (a) shows final configurations for the fraction of black (*b*) cells $f_b = 0.1$ to 0.5. (b) and (c) show sequences of configurations for $f_b = 0.3$ and 0.6. (d) shows the breakup of a line of cells into drops. See Appendix A for details.

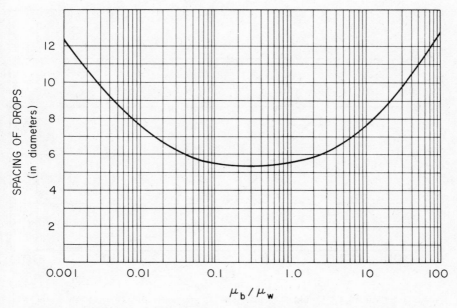

New Fig. 4. "Variation of disintegration wave length with viscosity ratio, as predicted by Tomotika's analysis." Reproduced from figure 1, Miesse (1955) with permission of the copyright owner, the American Chemical Society.

These calculations are based on a perturbation analysis in which it is "assumed that all the initial disturbances are very small and of the same magnitude" (Meister and Scheele, 1967). If there are any dents in the cylinder, they will bias the process toward different spacings. Trinkaus and Lentz (1964) observed a few elongated clusters, which could marginally be regarded as cylindrical, break up into two or three pieces after being formed by the fusion of small clusters. The longest such cluster, breaking into three pieces (Trinkaus and Lentz, 1964, figs. 29 and 30), had a spacing approximately four times its original diameter (new fig. 2). This finding is in plausible agreement with Tomotika's calculations if the viscosities of the two tissues are equal within an order of magnitude. With some ingenuity it should be possible to create a more perfect cylinder of one cell type embedded in an aggregate of another and to perform a more rigorous test.

Meister and Scheele (1967) give a graph (new fig. 5) of a correction function ϕ versus μ_b/μ_w in the equation

$$\alpha = \frac{\sigma_{bw}}{2c\mu_w}(1 - k^2c^2)\phi, \tag{B1}$$

from which it is possible to predict the rate at which the cylinder will break into drops. c is the cylinder radius, $k = 2\pi/\lambda$ is the wave number where λ is

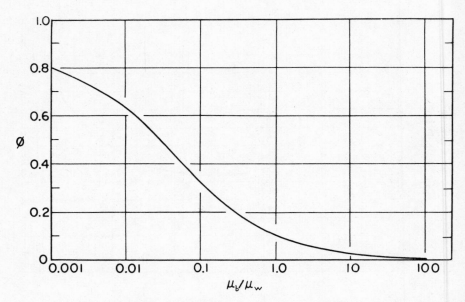

New Fig. 5. "Values of ϕ for use in equation [B1] to predict the disturbance growth rate for a high viscosity liquid jet in a high viscosity liquid medium." Figure 3 in Meister and Scheele (1967), reproduced with permission of Dr. Scheele.

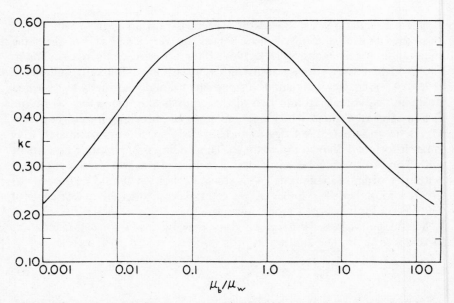

New Fig. 6. "Values of $[kc]$ which maximize the disturbance growth rate for a high viscosity liquid jet in a high viscosity liquid medium." Figure 2 in Meister and Scheele (1967), reproduced with permission of Dr. Scheele.

the spacing of the drops, and σ_{bw} is the interfacial tension. The perturbation creating the drops increases as $e^{\alpha t}$, where t is time. Thus, $1/\alpha$ is a characteristic time for formation of the drops. Trinkaus and Lentz' rough cylinder of cells discussed above had almost completely separated into three drops in a period of 52 hr (Trinkaus and Lentz, 1964, figs. 29 and 30). If $\mu = \mu_b = \mu_w$, then $\phi \approx 0.1$ (Meister and Scheele, 1967), and $kc = 0.563$ (new fig. 6). From this we calculate $\sigma_{bw}/\mu \simeq 2 \cdot 10^{-7}$ cm/sec. If (see Appendix D) $\sigma_{bw} = 1\text{--}30$ dyne/cm, then $\mu \simeq 5 \times 10^6\text{--}1.5 \times 10^8$ poise.

The following should be noted about the breaking of a cylinder into drops.

(1) Meister and Scheele (1967) give a criterion N_0^* for the applicability of Tomotika's analysis when μ_b/μ_w is between 0.01 and 10.0:

$$N_0^* = \frac{(2\sigma_{bw}d_b c)^{1/2}}{\mu_w + 3\mu_b}, \tag{B2}$$

where d_b is the density of the cylinder. Taking $d_b = 1$ g/cm^3, $c \simeq 1.3 \times 10^{-3}$ cm (Trinkaus, 1969, fig. 29), σ_{bw} as above, and $\mu_b = \mu_w = \mu$, then $N_0^* \simeq 10^{-7}\text{--}10^{-8}$. No significant departure from Tomotika's analysis occurs until N_0^* reaches 0.1. Thus the effects of inertia or momentum are entirely negligible in cell sorting.

(2) The breakup of a viscous cylinder into drops occurs in at least two other biological contexts—the beading of the diatom trail (Gordon and Drum, 1970) and drop formation from amoeba pseudopodia under high pressure (Marsland, 1956).

(3) One of the boundary conditions for solving the hydrodynamic equations for a cylinder of one fluid within another (Tomotika, 1935) is that "there is no slipping at the surface of the column" which "requires that the velocity components be continuous at the surface." This is a generally assumed condition in the solution of two-phase hydrodynamics problems. It may be violated by living cells because one sheet can slide over another. A question that must be asked is whether this involves large scale motion of cells past one another at a boundary. If slippage occurs, this additional parameter will have to be added for a full fluid mechanics description of cell sorting (Jacobson and Gordon, 1975).

(4) Another assumption we are making in applying hydrodynamical equations to cell populations is that elastic stresses generated by the interfacial tensions are relaxed in times short compared to sorting out. We are now testing this assumption (Phillips, Steinberg, and Lipton, 1972, and in preparation).

APPENDIX C

Rounding Up of an Ellipsoidal Aggregate into a Spherical One

In this appendix we will make an estimate of the tissue viscosity by comparing the tendency of rounding up of a roughly ellipsoidal aggregate into a nearly spherical one (see Steinberg, 1963) with rounding up of a fluid drop that is driven by the surface tension and resisted by the viscosity of the fluid.

Young (1939) has used an analogy between elastic and viscous bodies to put an elasticity theorem of Betti's into a form useful for calculating the rate of deformation of a viscous object that is subject to arbitrary volume and surface forces. The full equation could be used, for instance, to predict the rate of deformation of a sessile drop of reaggregated tissue in a centrifuge (Phillips and Steinberg, 1969). Young's result has been generalized to include objects with nonuniform viscosity (Isenberg, 1953; Rashevsky, 1960).

If an aggregate of cells forms a perfect ellipsoid of revolution generated by rotating the ellipse

$$z^2/a^2 + x^2/b^2 = 1 \tag{C1}$$

about the z axis, the ratio of the axes, $r = a/b$, ought to change according to

$$
\begin{aligned}
\frac{dr}{dt} &= \frac{1}{v^{1/3}} \frac{\sigma}{\mu} \frac{3}{8(r^2 - 1)} \left(\frac{4\pi r}{3}\right)^{1/3} \left(-2 - r^2 + \frac{r^2(4 - r^2)}{|r^2 - 1|^{1/2}} \right. \\
&\quad \times \left. \left\{ \begin{array}{l} \sin^{-1}[(r^2 - 1)^{1/2}/r] \\ \ln\,[(1 + (1 - r^2)^{1/2})/r] \end{array} \right\} \right) \\
&\equiv \frac{1}{v^{1/3}} \frac{\sigma}{\mu} \rho(r),
\end{aligned}
\tag{C2}
$$

where $v = \frac{4}{3}\pi ab^2$ is the volume of the ellipsoid (Young, 1939). The upper term in the curly brackets is taken when $r > 1$, the lower when $r < 1$. A plot of the function $\rho(r)$ is shown in figure 3.

If at time t_0 an ellipsoid of cells has an axial ratio r_0 and by time t_1 reaches the ratio r_1, then

$$
\begin{aligned}
\frac{\sigma}{\mu} &= \frac{v^{-1/3}}{t_1 - t_0} \int_{r_0}^{r_1} \frac{dr}{\rho(r)} \\
&\equiv \frac{v^{-1/3}}{t_1 - t_0} [\tau(r_1) - \tau(r_0)].
\end{aligned}
\tag{C3}
$$

The function $\tau(r)$, obtained by numerical integration, is plotted in figure 4.

It might be possible to form a prolate ellipsoid of revolution $(r > 1)$ (or even a cylinder) from a cell aggregate by using shear flow (Gopal, 1968) or by slowly pelleting cells into an appropriately shaped mold. Other plausible methods would be to slowly draw an aggregate through a tapering capillary (cf. Harvey and Shapiro, 1941) or to use a shaker with a linear stroke during reaggregation.

It is perhaps simpler to centrifuge reaggregating cells against the surface of a denser liquid, which would result in a flat lens shape. This could be regarded approximately as an oblate ellipsoid of revolution (oblate spheroid, $r < 1$).

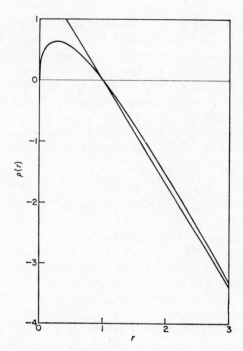

Fig. 3. The curved line is $p(r)$ in equation (C2). The straight line is Wheeler's approximation for $\mu_w/\mu_b = 0$ (equation (c4)).

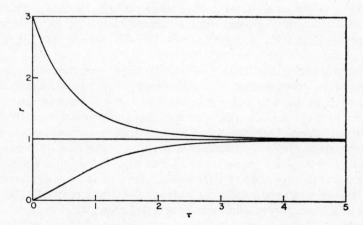

Fig. 4. r vs. the integral τ (effectively, dimensionless time), starting with $r = 0.01$ and $r = 3$ at $\tau = 0$.

Phillips and Steinberg (1969) found that certain cell aggregates $0.008 \times 0.04 \times 0.04 \, \text{cm}^3$ ($r = 0.2$) rounded up overnight (16 hr, H. Phillips, personal communication). If we assume that 'rounding up' corresponds to $r = 0.95$, then from figure 4 $\tau(0.95) - \tau(0.2) = 2.3 - 0.4 = 1.9$, so that $\sigma/\mu \approx 10^{-6}$ cm/sec [equation (C3)]. If $\sigma \simeq 1 - 30 \, \text{dyne/cm}$ (see section 4), $\mu = 1 \times 10^{6} - 3 \times 10^{7}$ poise. This result compares favorably with the estimates made in section 3 and Appendix B.

Wheeler (in Harvey and Shapiro, 1941) gives an approximate solution to the rate of rounding up of a viscous ellipsoid in a viscous medium, which in our notation becomes

$$\frac{dr}{dt} = \frac{1}{v^{1/3}} \frac{\sigma}{\mu_b} \left(\frac{4\pi}{3}\right)^{1/3} \frac{40(1 + \mu_w/\mu_b)(1 - r)}{(2 + 3\mu_w/\mu_b)(19 + 16\mu_w/\mu_b)}$$

$$\equiv \frac{1}{v^{1/3}} \frac{\sigma}{\mu_b} W(\mu_w/\mu_b)(1 - r), \qquad (C4)$$

where μ_b is the viscosity of the ellipsoid, and μ_w is the viscosity of the medium. $\mu_w/\mu_b = 0$ corresponds to an isolated ellipsoid and should be comparable to the solution of Betti's equation above [equation (C2)]. Indeed, for $r > 1$, the agreement is excellent (fig. 3).

Wheeler's approximation is equivalent to

$$r - 1 = \exp\left[-\frac{\sigma t}{v^{1/3}\mu_b} W(\mu_w/\mu_b) \right]. \qquad (C5)$$

Because $W(1):W(0) = 0.737:1.697 = 0.435$, an ellipsoid of b cells embedded in w cells of similar tissue viscosity will round up at a much slower rate than an isolated ellipsoid of b cells. This may explain why embedded clumps often apparently retain an ellipsoidal shape (see figures in Steinberg, 1962b, 1963, 1964, 1970), whereas isolated clumps become almost perfect spheres on a comparable time scale (Steinberg, 1963; Phillips and Steinberg, 1969) (new fig. 1).

An important part of Tomotika's (1935) result (see Appendix B) is that spacing of the drops formed from a viscous cylinder in a viscous medium is independent of the value of the interfacial tension σ. This outcome of course assumes that the surface and interfacial tensions are appropriate for the formation of drops. Only the rate of drop formation depends on σ. Similarly, for ellipsoids of revolution, the rate of rounding up is directly proportional to the surface tension σ [equation (C2)]. For sintering, dx/dt is again directly proportional to σ [equation (1)]. It would seem that there is a general lesson to be learned here: apparently, for a system consisting of two fluids, at least one of which is highly viscous, the sequence of geometrical configurations is independent of the value of the interfacial tension. σ affects only the time scale, a result that may be shown by dimensional analysis (Levich, 1962).

When three fluids are involved, the geometry additionally depends on the relative values of the interfacial tensions (Steinberg, 1963; Goel et al., 1970). This is the case for cell sorting when one asks which is the internally segregating phase, and for engulfing. However, when the external boundary of an aggregate may be ignored, sorting out of a pair of cell types may be regarded as a two-phase problem.

APPENDIX D

Experiments on the Rate of Cell Sorting versus Temperature

The following experiments on the temperature dependence of sorting of mixtures of chick embryonic cells from four-day limb precartilage and five-day liver were performed. Four-day limb buds and five-day livers were excised in Hanks' solution. Epidermis was removed from the intact limb buds by the cold trypsin method of Szabo (1955) (see also Rawles, 1963). The two sets of organs were then separately minced, rinsed in calcium- and magnesium-free (CMF) Hanks' solution, and each set was placed in 5 ml of a 0.1 % solution of Difco trypsin 1–250 dissolved in CMF Hanks' solution and contained in a 15 ml screw-capped, round-bottomed test tube. These tubes were warmed to 37°C in a water bath, placed at an angle of about 10° to the horizontal on a rotor, and rotated about their axes at 60 rev/min for 15 min at 37°C. The trypsin solution was then replaced with an equal volume of 10 % horse serum in Eagle's Minimal Essential Medium made up with Hanks' salts and containing 100 units of penicillin and 100 μg of streptomycin per ml. The contents of the tubes were then sheared with a Vortex mixer for 15 sec and centrifuged for 3 min at 189 g. The supernates were discarded, and the cell pellets were resuspended in 3 ml of the above medium and centrifuged at 40 g for 1 min to sediment large cell clusters. The supernates, containing mostly single cells, were transferred to fresh tubes and returned to the 37°C rotor for 2 hr (while adhesiveness returned). The resulting suspensions of cells and aggregated clusters were sheared with the Vortex mixer to break up the clusters, cell counts were made with a hemocytometer, and a mixture of 30 % limb bud cells and 70 % liver cells was centrifuged to form a thin pellet in a fresh round-bottomed tube. In order to avoid the possibility of density-dependent cell-type separation during centrifugation, the pellet was resuspended in a short column of medium (1 ml) and repelleted. This thin mixed pellet was incubated at 13°C for about 22 hr, during which time it became quite cohesive, although sorting out did not proceed to an extent detectable on histological sections. The pellet was then cut into many small pieces, which were divided among four 10-ml de Long flasks (Steinberg, 1962c), which were maintained upon gyratory shakers with $\frac{1}{2}$ in. diameter of gyration, at 120 gyrations/min, at 37.5°, 30°, 20°, and 10°C, respectively. From this point (taken as zero time), observations were made at intervals upon the living aggregates and upon fixed and stained material in serial histological sections.

The aggregates maintained at 10°C barely smoothed over initial gross surface irregularities even after 6.75 days. In fixed material, cells within such aggregates appeared loosely coherent. (But samples shifted to 37.5°C after 4 days at 10°C proceeded to sort out rapidly thereafter.) The aggregates maintained at 20°C were irregular and loose textured even after 2 days. At 4 days, their surfaces were still loosely textured although the interiors appeared to be tightening up. No sorting out was seen histologically even after 6.75 days. The aggregates maintained at 30°C showed substantial sorting at 24 hr, when the aggregates were roundish (ovals, for example) but not yet spheres. At 24 hr, the degree of sorting out was comparable to that observed in 37.5°C aggregates at ∼ 12 hr. At 48 hr, 30°C aggregates were approximately spherical and showed sorting comparable to that seen in 37.5°C aggregates at ∼ 32 hr. At 4 days, sorting out in the 30°C aggregates was comparable with that seen at 37.5°C at 2 days. The aggregates maintained at 37.5°C already showed a coarse patchwork of partially sorted areas at 6 hr in some cases but little sorting out in others. At 24 hr, sorting out was largely complete, but the liver–precartilage boundary was indistinct. By 48 hr, this boundary had become crisp.

APPENDIX E

The Interfacial Tension Between Unlike Tissues

In order to apply the viscous liquid model to the breakup of a cylinder into drops (Appendix B), we had to assume that the interfacial tension between unlike cell populations is of the same order of magnitude as that of a cell aggregate against culture medium. In this appendix we will try to justify this assumption.

When differing chick embryonic cell populations are brought together in pairs and allowed to rearrange, the configurations they approach are often found to be either a sphere of one tissue totally covered by a sphere of the other, or a spheroid of one tissue partially covered by a spheroid of the other (Steinberg, 1962b, 1963, 1964, 1970). If these configurations are determined by the adhesive free energies at the population interfaces, specific sets of adhesive energies are required to account for the various configurations. These adhesive free energies can be expressed either as works of adhesion, W (Steinberg, 1962, 1963, 1964), or as specific interfacial free energies σ (Phillips, 1969 and in preparation; Phillips and Steinberg, 1969). In the latter case, the adhesive relationships determining the total envelopment of population b

by w at configurational equilibrium are $0 < \sigma_{bw} \leqslant \sigma_{bo} - \sigma_{wo}$, whereas the relationships determining the partial envelopment of b by w are $\sigma_{bo} - \sigma_{wo} < \sigma_{bw} < \sigma_{bo} + \sigma_{wo}$. In recent experiments (Phillips and Steinberg, 1969), σ_{xo}, the surface tension of tissue x against the culture medium o, has been tentatively determined for several chick embryonic cell populations; the values appear to be in the range of 1–30 erg/cm^2 (1–30 dyne/cm if expressed as a surface tension) (Phillips and Steinberg, unpublished data). Assuming for populations b and w the values $\sigma_{bo} = 20$ and $\sigma_{wo} = 5$ dyne/cm, the range of values of σ_{bw} leading to either total or partial envelopment can be calculated from the above equations. For total envelopment, $0 < \sigma_{bw} < 15$ dyne/cm, whereas for partial envelopment $15 < \sigma_{bw} < 25$ dyne/cm. Because cases of both partial and complete envelopment at equilibrium have been observed, one would predict that σ_{bw} should often be of the same order of magnitude as σ_{bo} and σ_{wo}. A double sessile drop method of measuring σ_{bw} has been devised by Phillips, and a single measurement using one pair of tissues seems to support this prediction (Phillips, 1969) (new fig. 7).

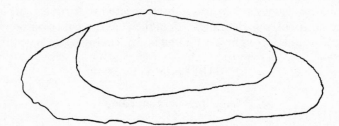

New Fig. 7. "Vertical section of an aggregate composed of initially intermixed heart and limb bud cells that sorted out while being centrifuged at 9000 g for 48 hours. The aggregate was fixed while being centrifuged. The limb bud cells have formed an island floating at the top of the heart cells that surround them. Maximum aggregate diameter is approximately 0.4 mm." Tracing from figure 22, Phillips (1969), with the author's permission.

One of us (Steinberg, 1964) has investigated the properties of a stochastic model for the adhesions between cells, in which a given cell type is characterized by a certain surface density of adhesive sites. This model yields the relation

$$\sigma_{bw} = (\sigma_b \sigma_w)^{1/2}. \tag{E1}$$

Because the distance between adhesive sites would vary from one cell to another in the stochastic model, they could not be close packed on all cell types. Thus, the stochastic model for adhesions is not consistent with our assumption in the viscous liquid model that the protrusions are both close packed and identical with the adhesion sites. However, a model is available for interfacial tensions of liquids, called the quasi-lattice model (Good and

Elbing, 1970), which makes the same assumptions we have made in our calculations of tissue viscosity in terms of the protrusions. It predicts

$$\sigma_{bw} = \sigma_b + \sigma_w - 2\Phi(\sigma_b\sigma_w)^{1/2}, \tag{E2}$$

where "Φ should be close to unity when the cohesive forces within each separate phase and the forces acting across the interface are of the same type." A more sophisticated theory (Good and Elbing, 1970) could not be applied until more is known about the chemistry of the protrusions. Both equations (E1) and (E2) give theoretical bases for the assumption that σ_{bw} is often of the same order of magnitude as σ_b and σ_w.

It is interesting to note that the stochastic model depends on quantitative differences in the number of adhesive sites per unit area, whereas the explanation of high tissue viscosities in terms of protrusions implies, under the assumptions of their close packing and their identity with adhesion sites, that the protrusions are different for different cell types. (Nevertheless all types of protrusions are assumed to interact, unlike the assumption in special hooks theories.) As an alternative to close packing it is possible that the protrusions might consist of ridges sufficiently interconnected to form a rigid network. The ridges on apposed membranes would then have to pass over one another. These effects could be roughly accounted for by leaving $\lambda_1 \neq \lambda$ in equation (2).

APPENDIX F

Engulfing and Spreading

The engulfing of one liquid drop by another immiscible drop suspended in a third immiscible liquid has been studied by Torza and Mason (1970). They found that "engulfing occurs by a combination of the simultaneous processes of (a) penetration and (b) spreading." The relative importance of the two processes depends on the drop sizes and interfacial tensions. A possible example of penetration with embryonic tissues may be seen in Steinberg (1970, fig. 25) (new fig. 8). This should be compared with the figures of Torza and Mason (1969, 1970) (new fig. 9).

The time course of the engulfing of two immiscible viscous liquids has apparently not yet been worked out mathematically. However, when spreading dominates the first stage, which seems to be the general case (Steinberg, (1970), we are dealing with fusion or sintering again, and equation (1) ought to be an applicable approximation, provided the interfacial tension between the unlike cell types is known (Appendix E). A rigorous mathematical derivation should include the line tension, which, however, is likely to be quite small (Torza and Mason, 1970) (cf. Torza and Mason, 1971).

Alternatively, if the tissue viscosities μ_b and μ_w of tissue types b and w were known, equation (1), or its exact formulation for unlike liquids, could be used to measure σ_{bw}. This would be a kinetic alternative to the equilibrium double sessile drop method (Phillips, 1969).

It is of interest to note that phagocytosis has been quantitatively shown to be a phenomenon of engulfing driven by interfacial tensions (van Oss and Gillman, 1972). Apparently the engulfing step of phagocytosis is a passive process because when the surface tensions are equal, bacteria simply adhere to white blood cells and are neither engulfed nor lysed (van Oss, personal communication).

Similarly, cell adhesion to flat substrates may be an attempted engulfing (van Oss, personal communication). The spreading of a drop on a flat surface, driven by surface tensions, can be rate limited by either the rate of wetting of the substrate at the edge or by the high viscosity of the drop (Dettre and Johnson, 1970). If wetting is rate limiting, the drop retains the shape of a spherical segment. An absolute rate theory, analogous to that used for bulk viscosity (Glasstone et al., 1941), has been developed to explain such spreading (Cherry and Holmes, 1969). If the resistance of the drop to change of shape, namely its viscosity, is rate limiting, the drop tends to be bell shaped (Dettre and Johnson, 1970). Another example of this may be seen in Torza and Mason (1969, 1970) (new fig. 9). Perhaps these considerations could be applied to the spreading of cell sheets, such as epiboly in teleosts (Trinkaus, 1969).

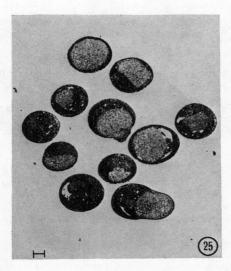

New Fig. 8. "Section through 11 of the 12 heart-liver fusion masses comprising one experiment. Liver (darkly stained) has covered the heart ventricle fragment completely in some cases and partially in others. Two and three-quarter day culture." Bar, 0.1 mm. Figure 25 from Steinberg (1970), with the author's permission.

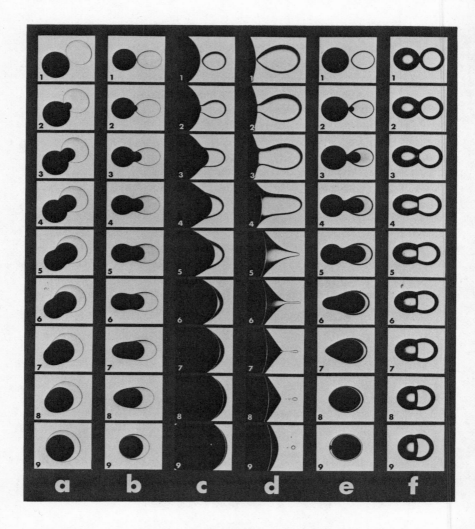

New Fig. 9. "Photographs of engulfing steps in system 6 (*sequences a, b, c, d*) when shear with counter-clockwise rotation (*a*) or horizontally applied electrical (*b, c, d*) fields caused the two drops to approach one another; and for system 20(*e*) and system 13(*f*) in a horizontal electrical field. In all sequences phase-1 is dark. Penetration of phase-1 into phase 3 became less pronounced from sequence *b* ($\Delta p_{13}^0 = 65$ dynes cm^{-2}) to sequence *c* ($\Delta p_{13}^0 = 28$ dynes cm^{-2}) because of the decrease in Δp_{13}^0 and disappeared completely in sequence *d* ($\Delta p_{13}^0 = -9.7$ dynes cm^{-2}).... Partial coalescence can be observed in frames 6 to 9 of sequence *d* .... The total time intervals of the sequences were about 1 sec for *a, b, e*; 3 sec for *f*; and 5 sec for *c* and *d*, where the engulfing was dominated by the spreading of the relatively viscous phase-3. The frames were selected from cinephotographs taken at 600 pictures per sec." $\Delta p_{13}^0 = 2\sigma_{12}/r_1 - 2\sigma_{23}/r_2$ is the "capillary pressure difference," where r_1 and r_3 are the initial radii of the drops, and subscript 2 refers to the surrounding medium. See Torza and Mason (1970) for the liquids used in each "system." Reproduced with permission of Dr. Torza.

12. Transitivity, Pattern Reversal, Engulfment, and Duality in Exchange-Type Cell Aggregation Kinetics

P. Antonelli, D. I. McLaren, T. D. Rogers,
M. Lathrop, and M. A. Willard

Simulations of cell-sorting phenomena in embryogenesis describable by means of the differential adhesion mechanism are further studied by use of the exchange model. In this study, more than two cell types are employed and results nicely complement the successes of earlier work with only two cell types. Again, it is noticed that central clumping does not usually appear in final distributions. In addition, the phenomena of transitivity, pattern-reversal, and the engulfment of one tissue by another are studied. The first is shown to be a trivial consequence of Steinberg's stochastic version of differential adhesion. Also, it is shown that the exchange model successfully simulates pattern reversal by reversing the relative adhesion values due to the time-in-culture reversal of the adhesivity effect of trypsin-assisted dissociation of tissues. The connection between the occurrence of pattern reversal and the absolute and relative positions of tissue types in the adhesion hierarchy is explained in terms of the exchange model. However, it is found that the model, as it stands, does not seem to give a satisfactory basis for the simulation of engulfment. Perhaps our suggestions for the extension of the exchange principle will rectify this limitation. We maintain that any real system that exhibits engulfment will also exhibit central clumping, and conversely. The importance of the figure–ground ambiguity that we have called duality is again stressed. If this ambiguity is not taken into account in experimental work, interpretations of many results on cell sorting are rendered meaningless.

1. INTRODUCTION

In a previous paper (Antonelli et al., 1973, hereafter referred to as ARW), Steinberg's differential adhesion hypothesis for the theory of cell sorting was distilled to an operational recipe called the exchange principle, which has as essential ingredients: (1) differential adhesion among cells, (2) hexagonal cell geometry, and (3) certain simple thermodynamic considerations such as the minimization of free energy. The exchange principle allows or disallows the exchange of a given (randomly chosen) cell in an aggregate with one of its nearest neighbors, according to whether or not the free energy is thereby decreased. The simple form of the exchange principle makes possible the prediction of local behavior by inspection and is readily adapted to provide simulations of cell sorting on a digital computer.

The development of the exchange principle in ARW was supported by simulations involving two cell types and gave an accurate prediction of the time course of cell sorting as observed experimentally by Trinkaus and Lentz (1964). Further support for the exchange principle and therefore for Steinberg's theory was presented in arguments advanced against experimental (Roth and Weston, 1967) and theoretical (Goel et al., 1970) positions that purported to prove the inadequacy of the Steinberg theory. A prediction arising from the simulations presented in ARW was the following: "The large-scale geometry of the stable state can be seen in the large-scale geometry of the initial configuration in all cases." This conclusion, which on the one hand reflected the inability of our model to produce Steinberg's centrally clumped stable states from an initially random distribution, contrasted on the other hand with its outstanding success in reproducing Trinkaus and Lentz's (1964) markedly different final distributions. The simulations involving two cell types (one of which might be regarded as the medium) also gave rise to a phenomenon akin to the problem of figure–ground ambiguities found, for example, in studies on visual perception, and that we called *duality* in ARW. More will be said about this later in this section and in section 5 of the present paper.

We report here the results of using more than two cell types, and we present a study of three phenomena associated with cell-sorting processes, namely transitivity, pattern reversal, and engulfment. These matters are briefly introduced in the remaining part of this section, which precedes the discussion of three- and four-cell-type simulations. A detailed discussion of each phenomenon follows in sections 2, 3, and 4, respectively.

Transitivity. Suppose that in experiments using two cell types it is found that type 1 segregates internally to type 2, and type 2 segregates internally to type 3. If it is subsequently found that type 1 segregates internally to type 3, the cell sorting of these three tissue types is said to possess the property of *transitivity*. Expressed symbolically for a relation R, transitivity is

$$aRb \quad \text{and} \quad bRc \Rightarrow aRc.$$

Transitivity in cell sorting and its relationship with the notion of a hierarchy of adhesive strengths among cell types have been discussed extensively by Steinberg (1970). The purpose of raising it here is to demonstrate (and provide an easy remedy for the fact) that transitivity is not a consequence of the exchange principle.

Pattern Reversal. Ideally speaking (see, for example, Steinberg, 1964 or Trinkaus, 1969), given two cell types 1 and 2, the type that segregates internally from an initially random aggregate of type-1 and type-2 cells should reflect only the type with the stronger self-adhesion and should be independent of the relative proportions (the phase ratio) of the two cell types present. In general, this is experimentally true but not always so : Using chick embryonic heart and liver cells, Wiseman et al. (1972) have made a systematic study of the sorting out of heterotypic reaggregates, of the fusion of isotypic re-aggregates, and of the fusion of tissue fragments through experiments where the type of the internally segregating component depended on the relative quantities of heart and liver cells involved. For a range of phase ratios, a reversal of the phase ratio from liver dominant to heart dominant (before sorting out or reaggregate fusion) resulted in a switch of the cell type identi-fiable as the externally segregated component at the end of sorting (or fusion). It is natural that we should attempt to simulate this phenomenon (for hetero-typic reaggregates) by use of the exchange model.

Engulfment. As mentioned above, Wiseman et al. (1972) report, in addition to experiments on the sorting out of reaggregates of mixed cell types, the results of a different type of experiment, wherein a study of the fusion behavior of (1) intact (precultured) tissue fragments of differing types, and (2) reag-gregates of differing types is made. In the case of fusion of reaggregates it was shown that pattern reversal (discussed above) could be obtained from a variation of the relative sizes of the two reaggregates. What is initially of interest to us is the fact that one of the two components in the fusion experi-ments proceeds to surround and may even succeed in surrounding the other component. We call this phenomenon *engulfment*. Attempted simulations of engulfment and an analysis of the strategies involved lead to the conclusion that the exchange model cannot simulate engulfment and that modifications are therefore necessary.

Simulations with 3 or 4 Cell Types. Regarding the results of simulations employing 3 and 4 cell types, it is correct to say that the general qualitative findings for 2 cell types (particularly with respect to the time sequence of sorting out) also hold for 3 and 4 cell types and may reasonably be expected to hold for any number of cell types. The results of simulations with 3 types are shown in figures 1, 2, 4, and 5. It was expected that the use of 3 cell types (one of which might be regarded as representing the culture medium) would allow a more realistic simulation of the sorting out of two real cell types, particularly when dealing with the phenomenon of pattern reversal. With simulations of 2 cell types, the manifestation of duality meant that pattern reversal was a necessary consequence of phase reversal, hardly a suitable starting point for simulations of its occasional appearance. With 3 cell types, except when the proportion of the medium type is very low, there exists a figure–ground ambiguity in the interpretation of the final distributions. This ambiguity means there is a problem when it comes to deciding which of two components (embedded in a sea of the third medium component) can be said to surround or segregate externally to the other. In lieu of a rigorous solution to this problem and in order to be able to interpret the simulations, it was necessary to undertake a diachronic study of the simulations. That is, the *local* time evolution of, and cell movements within, the simulated re-aggregates were studied to decide which cell type ultimately surrounded which.

2. TRANSITIVITY

It will be shown here that transitivity is not a consequence of the exchange principle alone and that it is necessary to invoke the ideas embodied in Steinberg's stochastic model (see Steinberg 1964, 1970), as modified by geometric considerations, in Appendix B of ARW. The stochastic model is an alternative view of certain aspects of the differential adhesion theory.

As discussed in ARW, we suppose the force of adhesion (essentially the negative of the free energy change) between cell type i and cell type j is represented by the nonnegative number λ_{ij}. Then, if we consider a random aggregate made up of cells of types i and j, the exchange principle leads to the following rules:

If it is usually true in a given simulation that

$$\lambda_{ii} + \lambda_{jj} > 2\lambda_{ij}, \tag{1}$$

sorting out will occur. If it is usually true that

$$\lambda_{ii} + \lambda_{jj} < 2\lambda_{ij}, \tag{2}$$

mixing of the simulated tissues will occur.

Consider now aggregates of (a) cell type 1 with cell type 2, (b) cell type 2 with cell type 3, and (c) cell type 1 with cell type 3. Suppose that type 2 segregates internally to type 1 and externally to type 3, respectively. That is,

$$2\lambda_{12} < \lambda_{11} + \lambda_{22} \quad \text{and} \quad 2\lambda_{23} < \lambda_{22} + \lambda_{33} \tag{3}$$

with

$$\lambda_{11} < \lambda_{22} < \lambda_{33}. \tag{4}$$

It does *not* follow from relations (3) and (4) that

$$2\lambda_{13} < \lambda_{11} + \lambda_{33}. \tag{5}$$

Thus, the exchange principle does not predict (or does it deny) transitivity among these 3 cell types. To resolve this problem it is necessary to take a closer look at what is going on, along the lines indicated in Appendix B of ARW.

The central tenet of the stochastic model is that the force of adhesion between two cells, types i and j, is proportional to the product of the densities d_i and d_j of adhesion sites on the respective cell surfaces. It therefore seems appropriate for simulation purposes to write

$$\lambda_{ij} = d_i d_j. \tag{6}$$

From equation (6) it follows trivially that

$$\lambda_{ij} = (\lambda_{ii}\lambda_{jj})^{1/2}. \tag{7}$$

Now the force of adhesion between two cells depends critically on the amount of apposition between their surfaces. Because we have assumed maximal packing of cells, this apposition factor is uniform throughout the aggregate and so does not appear in equation (6). However, it is still possible that equation (7) should be replaced by

$$\lambda_{ij} = k_{ij}(\lambda_{ii}\lambda_{jj})^{1/2}, \tag{8}$$

where $0 < k_{ij} \leqslant 1$. The newly incorporated k_{ij} reflects the view that maximal apposition of surfaces may not be utilized. In any case, using either equation (7) or (8) we obtain

$$\lambda_{ij} \leqslant \frac{\lambda_{ii} + \lambda_{jj}}{2} \tag{9}$$

because the geometric mean is less than or equal to the arithmetic mean, that is,

$$\sqrt{\lambda_{ii}\lambda_{jj}} \leqslant \frac{\lambda_{ii} + \lambda_{jj}}{2}. \tag{10}$$

Therefore, transitivity is automatically satisfied in the stochastic version of the differential adhesion theory. Consequently, segregation between cell types will always occur in a random mixture.

It is worth noting that equation (8) is in accord with Parsegian and Gingell's (1972) comments arising from their calculations of the adhesive forces between idealized bilayer membranes. In particular, the repulsive (mixing) force due to the coulomb force between negatively charged acidic groups in the cell coats is found to be proportional to the product of the charge densities in the respective cell coats; the latter may be likened to the densities of adhesive sites in the stochastic model. Finally, inspection of the simulations shown in figures 1, 2, 4, and 5 in conjunction with satisfaction of the inequalities of type (1) among the λ values indicates that transitivity is successfully simulated.

3. PATTERN REVERSAL

In further experiments designed to reveal the causes of the pattern-reversal phenomenon, Wiseman et al. (1972) discovered that heart-cell adhesivity (1) increases with time in culture and (2) is decreased markedly in the course of the dissociation–reaggregation procedure employed in their cell-sorting experiments. The latter effect was attributed to the use of trypsin as a dissociation agent. To date, there is no published report of their proposed experiments to determine whether similar effects hold for the liver cells as well. In short, heart and liver isotypic reaggregates (also heart and liver in heart–liver reaggregates) start out in the sorting process with their positions in the adhesion hierarchy temporarily reversed, that is, with heart cells less adhesive than liver cells, and so cells of the heart component begin to move to the "external" position. However, as time progresses, heart adhesiveness increases up to and beyond that of liver cells, and liver cells tend to the external position. Ultimately, the component perceived to be externally segregated and that perceived to be internally segregated are dependent on the phase ratio of the two cell types. If the heart–liver ratio is sufficiently high, by the time the adhesion reversal takes effect the heart cells will have established an irreversible stand as the externally segregated component. The reader should consult Wiseman et al. (1972) for a detailed discussion.

Clearly, this is an observation for which the exchange model should be used to attempt a simulation. Rather than allowing for a continuous time-variation of the force-of-adhesion parameters λ_{ij}, it was considered sufficient that at some intermediate time in the simulation the λ values be appropriately changed (subject to certain constraints discussed below) and the simulation allowed to continue to its completion.

For a given choice of the λ values and of the time variation thereof, there are two simulations to be carried out: one with the phase ratio $x\%:(100 - m - x)\%$ and one with the phase ratio $(100 - m - x)\%:x\%$ of type-2 cells (liver) to type-3 cells (heart). where the proportion of type-1 (medium) cells is $m\%$. The results of such a set of simulations, with $m = 50$ and $x = 15$, are shown in figures 1–6. Simulations employing the same λ values and the same relative proportions of type-2 cells to type-3 cells, but with a substantially smaller proportion of medium cells ($m = 20, x = 25$), yield the same qualitative conclusions. The simulations displayed here are marginally easier to interpret than those of the latter case.

Figs. 1–6. Simulations of sorting with three cell types [1 (undrawn), 2 (white), and 3 (black)] showing sorting out and pattern reversal. In figures 1–3 the phase ratio is 50:15:35 of type 1:type 2:type 3; in figures 4–6 it is 50:35:15. Figures 1 and 4 show initial distributions, figures 2 and 5 are distributions at time unit 5 and figures 3 and 6 at time unit 10 (with changed λ-values). The λ values up to and including time unit 5 are

$$\lambda_{11} = 0 \quad \lambda_{12} = 0 \quad \lambda_{13} = 0$$
$$\lambda_{21} = 0 \quad \lambda_{22} = 0.8 \quad \lambda_{23} = 0.55$$
$$\lambda_{31} = 0 \quad \lambda_{32} = 0.55 \quad \lambda_{33} = 0.5$$

and after time unit 5

$$\lambda_{22} = 0.6 \quad \lambda_{33} = 0.7.$$

Fig. 1.

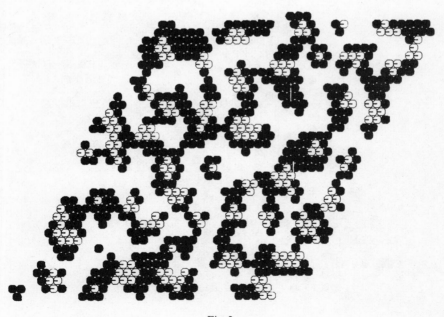

Fig. 2.

Fig. 3.

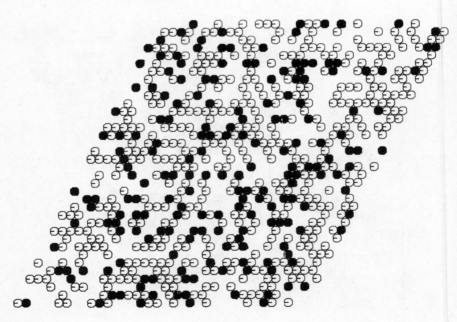

Fig. 4.

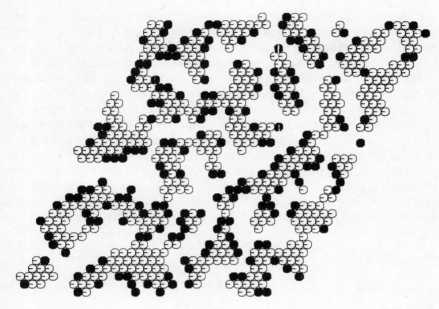

Fig. 5.

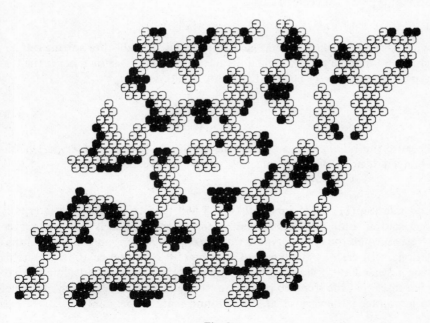

Fig. 6.

Comparison of the stable states and inspection of the respective preceding states indicate that the phase reversal has indeed led to pattern reversal. Note that until the adhesion reversal has taken place, the trend is for the less adhesive type-3 (heart) cells to segregate externally regardless of the phase ratio. After the adhesion change there is a reversal of roles, with type-2 cells moving out of type-3 cell clusters, and when the type-3 cell : type-2 cell ratio is small (15 : 35), the type-2 cells succeed in becoming the externally segregated component. For the other ratio (35 : 15), the type-3 cells retain (through weight of numbers) their external position, and it is the existence of this abnormal stable state that constitutes pattern reversal.

We now consider these simulations from the point of view of the exchange principle. There are 3 cell types, one of which (type 1) is regarded as having zero or relatively very low adhesion with itself and the other types. If sorting out occurs with types 2 and 3, with type 2 segregating internally, then

$$\lambda_{23}\,(=\lambda_{32}) < \frac{\lambda_{22} + \lambda_{33}}{2} \tag{11}$$

on the average, and also

$$\lambda_{33} < \lambda_{22} \tag{12}$$

so that type 2 is internal. Taking

$$\lambda_{33} < \lambda_{23} \tag{13}$$

with relation (11) ensures that inequality (12) will hold. After sorting out has gone on for some time (5 time units in the simulations shown here), the iso-typic λ-values may be changed as follows (with $\varepsilon > 0$):

$$\lambda'_{22} = \lambda_{22} - \varepsilon$$
$$\lambda'_{33} = \lambda_{33} + \varepsilon \tag{14}$$

because the *relative* values determine sorting behavior, and λ_{23} is left unchanged. If now, instead of inequality (12) we have

$$\lambda'_{22} < \lambda'_{33}, \tag{15}$$

and relation (11) still holds, cell type 3 will tend to become the internally segregating component (as evident in the diagrams) and is finally observed to be so when the type 2 : type 3 ratio takes the larger value. Further simulations with various λ values and changes (14) such that relation (15) holds true revealed the following fact: changes (14) alone are not sufficient to guarantee that the trend to pattern reversal is stable or permanent. In order to guarantee pattern reversal, it was found that the inequality

$$\lambda'_{23} = \lambda_{23} < \min\{\lambda'_{22}, \lambda'_{33}\} \tag{16}$$

must also be obeyed. This requirement is a reflection of the conclusion (previously quoted) reached in ARW that the geometry of the final state is similar to that of the initial state; in the present case the initial state is the partially sorted state achieved after 5 time units.

We conclude from the model that pattern reversal results from a change in the λ values that favors all isotypic adhesions over heterotypic adhesions. Furthermore, it is not surprising that pattern reversal is observed in only one of five tissue-type combinations studied experimentally by Wiseman et al. (1972).

The occurrence of this phenomenon is determined by three factors: (1) the relative and (2) the absolute positions of the various types in the adhesion hierarchy, and (3) the direction and degree of the experimental perturbations that change the adhesive strengths from their "natural" values in the first place. Assuming the latter to be more or less constant, the occurrence of the strength reversal [relations (12)–(15)] will depend on the degree of proximity of the adhesivenesses of the different types in the hierarchy and on whether each type has a strong or mild preference for isotypic over heterotypic bonds. Here we take mild to mean

$$2\lambda_{ij} < \lambda_{ii} + \lambda_{jj},$$

and strong to mean

$$\lambda_{ij} < \min \{\lambda_{ii}, \lambda_{jj}\}.$$

The latter is true (on the average) if

$$2\lambda_{ij} \ll (\lambda_{ii} + \lambda_{jj}).$$

If the isotypic preferences are strong and/or the hierarchical ranking of the types is very close, it is likely that pattern reversal will be observed in experiments employing these cell types.

4. FUSION AND ENGULFMENT

Unfortunately, the end result of a considerable variety of simulations was the conclusion that the exchange principle model seemed unable to provide a simulation of (even partial) engulfment. Figures 7–9 show excerpts from the time course of a typical attempt, where the hope was that the less adhesive white cells would move to surround the group of black cells. The state shown for time unit 7 is stable. A glance at the relevant inequalities among the λ values will indicate the strategy behind this attempt. The inequalities

$$\lambda_{12} < \tfrac{1}{2}(\lambda_{11} + \lambda_{22}) \tag{17a}$$

and

$$\lambda_{12} < \tfrac{1}{2}(\lambda_{11} + \lambda_{33}) \tag{17b}$$

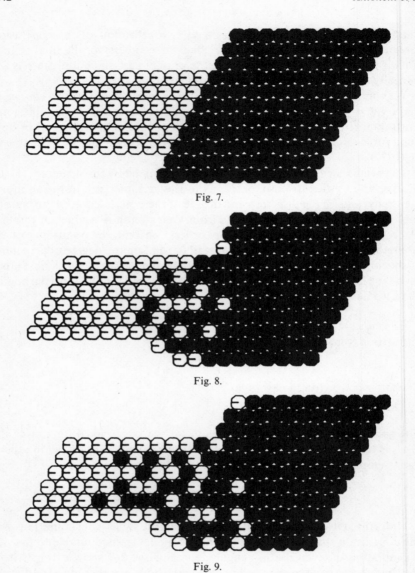

Fig. 7.

Fig. 8.

Fig. 9.

Figs. 7–9. Attempted simulation of engulfment. The λ values used are

$$\lambda_{11} = 0.2 \qquad \lambda_{12} = 0.15 \qquad \lambda_{13} = 0.1$$

$$\lambda_{21} = 0.15 \qquad \lambda_{22} = 0.6 \qquad \lambda_{23} = 0.75$$

$$\lambda_{31} = 0.1 \qquad \lambda_{32} = 0.75 \qquad \lambda_{33} = 0.8.$$

Figure 7 shows the initial distribution and figures 8 and 9 the distributions at time units 3 and 7, respectively.

are designed to ensure that cell types 2 and 3 do not mix with the medium cells of type 1. The mixing inequality

$$\lambda_{23} > \tfrac{1}{2}(\lambda_{22} + \lambda_{33}) \tag{17c}$$

ensures a degree of relative mobility of types 2 and 3, whereas

$$\lambda_{22} < \lambda_{23} < \lambda_{33}, \tag{17d}$$

it is hoped, will help to counteract the mixing effects of inequality (17c) by encouraging a preference for isotypic (3–3) bonds over heterotypic (2–3) bonds. However, the mixing tendency dominates and the strategy fails.

A closer examination of a different strategy will provide clues as to why the exchange principle is inadequate and will lead to some ideas as to how it should be extended or modified. Suppose we have (see fig. 10) a group of

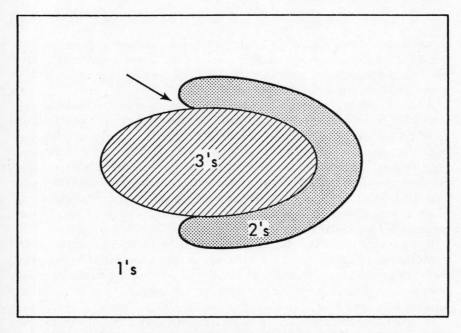

Fig. 10. A possible state of a system in the process of engulfment. Type-3 cells are hatched, type-2 stippled, and type-1 cells represent the surrounding medium.

type-3 cells partially surrounded by a group of (less adhesive) type-2 cells in a surrounding medium of type-1 cells. Focusing on the region indicated by the arrow in figure 10, we study the motion of a leading edge of the group of type-2 cells as it continues to engulf the type-3 cells. As with the preceding strategy, we require that inequalities (17a) and (17b) be satisfied in order to preserve the integrity of the groups of type-2 and type-3 cells, respectively, and to prevent them from mixing with the medium. In contrast to (17c) we require in addition that types 2 and 3 do not mix with each other:

$$\lambda_{23} < \tfrac{1}{2}(\lambda_{22} + \lambda_{33}). \tag{18}$$

The arrowed region of figure 10 is depicted in figure 11(a). Using the exchange principle, we discover the consequences of allowing three possible exchanges [depicted in figs. 11(b), (c), and (d)] that may reasonably be considered as contributing to further progress in engulfment. In all three cases, if the exchange is to be energetically allowed, it is found that the inequality

$$\lambda_{12} > \tfrac{1}{2}(\lambda_{11} + \lambda_{22}) \tag{19}$$

must be obeyed, in contradiction to inequality (17a). The weakest substitute possible for (17a) is the equality

$$\lambda_{12} = \tfrac{1}{2}(\lambda_{11} + \lambda_{22}), \tag{20}$$

in which case moves (b) and (c) would be allowed as zero-energy moves (see ARW) or as random walks. Such a passive mechanism seems unlikely to produce engulfment, let alone the ultimate rounding up that characterizes many of the experiments (Wiseman et al., 1972).

Next we consider a fluctuating strategy, whereby some of the time inequality (17a) holds and otherwise relation (19) or (20) is true. The idea is to allow type-2 cells to wander off (a little) into the type-1 cell sea, (perhaps) later to reaccrete to the main body of type-2 and type-3 cells or to allow progressive moves that are otherwise blocked. One consequence of this strategy is a steady loss of type-2 cells into the medium; simulations indicated this loss and the failure of this strategy to effect any progress toward engulfment. Furthermore, it would not be easy to explain and experimentally verify the meaning of these fluctuations in terms of the properties of real cells.

It does seem, therefore, that a point has been reached at which some modification or extension of the exchange principle is needed in order to facilitate simulations of engulfment, for it must be assumed that differential adhesion, as described by the exchange principle alone, is insufficient to account for engulfment.

In order to simulate pseudopodal extensions and explorations by the (advancing) type-2 cells, the following extended motility rules may be proposed for those type-2 cells with (a) at least one type 1 cell contact or (b) one or (more likely) two type-3 cell contacts and at least one type-1 cell contact. For case (a), inequality (17a) may be reversed as long as the moves thereby allowed retain at least one type-2 cell contact or establish at least one type-3 cell contact. A type-2 cell in case (b) may be allowed to break a type-2 cell contact, provided that it retains its type-3 cell contact(s). Assuming that progressive moves at the advancing edge of the type-2 cell group can be simu-

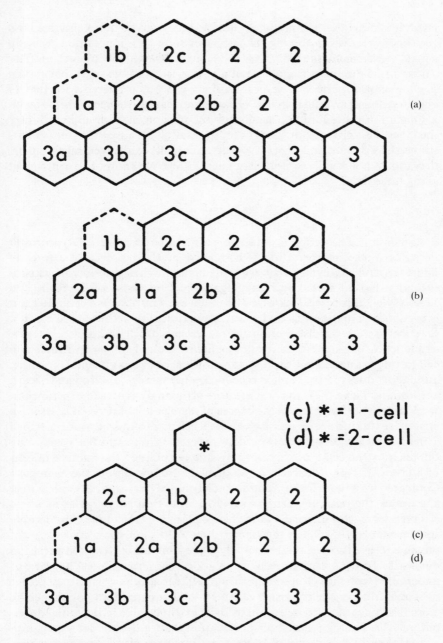

Fig. 11a–d. Closeup views of the arrowed region of figure 10 showing three possible movement strategies. Cases (c) and (d) differ according to whether the cell marked * is a type-1 or a type-2 cell.

lated, it is clear that the modified motility rules will be more effective than zero-energy moves in forcing the necessary follow-up moves that result in a mass movement of the type-2 cell body. Simulations involving such motility rules will be the subject of a later study. It is felt that the central clumping phenomenon reported by several workers can be correlated with that of engulfment and may therefore have a common causative mechanism. Hence, a simulation procedure that facilitates one phenomenon should encourage the other. This proposition leads us to suggest that because the tissue systems studied by Trinkaus and Lentz (1964) do not usually exhibit central clumping, they should not usually exhibit engulfment either. We know of no experiment along these lines.

5. DUALITY

As mentioned in section 1, simulations with three (or more) cell types result in states whose interpretation is hampered by the presence of a form of figure–ground ambiguity that we call duality. When considering experimentally derived sorted-out reaggregates such as those of Trinkaus and Lentz (1964), Elton and Tickle (1971), and figure 4 of Wiseman et al. (1972), it is even more important to find a way around duality because here it is not practical to study the distribution and movements of cells in a given aggregate as a function of time. The existence of a figure–ground ambiguity problem in cell sorting has never been explicitly acknowledged in experimental literature, although it does receive implicit recognition in the work of Elton and Tickle.

Elton and Tickle (1971) have as a feature of the analysis of their experimental work a statistical measure of the degree of segregation of two cell types in an aggregate. Taking aggregates that have, according to this measure, attained a marked degree of segregation, they proceed to measure the positioning obtained by the cell types involved by simply counting the number of cells of a given type per square in a rectangular grid laid across the aggregate. Compared with those for the randomly distributed (zero segregation) control aggregates, the resulting histograms allow fairly firm judgment as to which cell type segregates internally and which externally. Due to the considerably smaller number of cells in the standard 40×40 simulation grids, it is not possible to use this technique to analyze the simulations featured earlier in our work. To try it out, a special simulation based on a 50×120 grid was undertaken with the cell count taken across the longer axis. The result is that the count of type-2 cells in each of ten 11×11 squares does not change by more than one standard deviation from its original value in the initial distribution. This finding reflects our observation that in this model segregation is a local and not a global phenomenon. (For example, it does not exhibit central clumping.) Because Elton and Tickle's analysis of their experiments does reveal a significant variation in the distribution of cell types that show movement of heart cells toward the outside of the aggregate (which may be viewed as a tendency to central clumping), it is clear that some extension of the exchange model is called for. It seems likely that extended motility rules similar to those suggested for engulfment will be required.

It is easy to underestimate the full significance of the figure–ground ambiguity with regard to cell-sorting experiments. We emphasize again that, without a tried and tested quantitative measure, it is very difficult in many cases to judge a cross section of an aggregate by inspection and to state with certainty which cell type has segregated externally or internally.

6. Conclusions

Simulations based on the exchange model that employ more than two cell types essentially mimic the behavior and successes of earlier work (ARW) with two cell types. The limitations of the exchange model regarding simulations of engulfment and central clumping are discussed, and the importance of the figure–ground ambiguity that we call duality is stressed further. If duality is not properly taken into account, the interpretation of many experimental situations is rendered meaningless. An analysis of transitivity from the point of view of the modified (ARW) Steinberg stochastic model reveals that transitivity is a trivial consequence of the stochastic model and that mixing occurs only as a concomitant of sorting out. The exchange model successfully simulates pattern reversal by reversing relative adhesion values due to time-in-culture reversal of the adhesion effect of trypsin-assisted dissociation of tissues. The connection between the occurrence of pattern reversal and the absolute and relative positions of tissue types in the adhesion hierarchy is explained in terms of the exchange model. It is found that the exchange principle alone does not provide sufficient foundation for simulation of the engulfment of one group of cells by another contiguous group; alterations of motility rules (which extend the exchange principle) akin to pseudopod formation are suggested for further study. Because we feel that the phenomena of central clumping and engulfment are closely linked, these modifications are expected to facilitate central clumping too.

We would like to thank L. Wolpert, C. Tickle, and D. Garrod for their suggestions and encouragement regarding this work. Partial support for this research was received from NRC Grants A-8014 and A-5210.

REFERENCES

Antonelli, P., Rogers, T. D., and Willard, M. (1973) *J. Theor. Biol.* 41: 1–21.
Elton, R. A. and Tickle, C. (1971) *J. Embryol. Exp. Morph.* 26: 135.

Goel, N. S., Campbell, R. D., Gordon, R., Rosen, R., Martinez, H., and Yčas, M. (1970) *J. Theor. Biol.* 28: 423.

Parsegian, N. and Gingell, D. (1972) *J. Adhesion* 4: 283.

Roth, S. A. and Weston, J. A. (1967) *Proc. Nat. Acad. Sci. U.S.* 58: 974.

Steinberg, M. A. (1964) The problem of adhesive selectivity in cellular interactions. 22nd Symposium of the Society for the Study of Growth and Development. In *Cellular Membranes in Development*, ed. M. Locke. pp. 321–66. New York: Academic Press.

Steinberg, M. S. (1970) *J. Exp. Zool.* 174: 395.

Trinkaus, J. P. and Lentz, J. P. (1964) *Develop. Biol.* 9: 115.

Trinkaus, J. P. (1969) *Cells Into Organs*. Englewood Cliffs, N.J.: Prentice-Hall.

Wiseman, L. L., Steinberg, M. S., and Phillips, H. M. (1972) *Develop. Biol.* 28: 498.

Annotated Bibliography

Richard D. Campbell

The following list of publications, arranged alphabetically by author, is intended to provide a compact summary of current modeling papers dealing with cell movement. The only papers included are those that contain mathematical descriptions or predictive models of some type of cell translocation. Not all such papers have been included, but it is hoped that the selections provide a representative overview of this interesting area of biology.

Assheton, R. (1910) The geometrical relation of the nuclei in an invaginating gastrula (e.g., amphioxus) considered in connection with cell rhythm, and Driesch's conception of entelechy. *Wilhelm Roux' Arch. Entwicklungsmech. Organ.* 29:46–78.

This paper, which presents a geometrical theory of embryonic gastrulation, is representative of a number of early papers on cell movement. The initial force for invagination is considered to reside in a tension among cell centers, presumably mediated by cell–cell contacts. The pattern of invagination is determined by the positions of cell contacts (and hence of force transmittal) on the surface of each cell. If, on one side of the blastula, the cells tend to have their contacts external to a spherical shell defined by the centers of all cells, and on the other side of the blastula the contacts are internal to this shell, the former side should buckle into the latter. This paper is explicitly aimed at refuting Driesch's theory of entelechy.

Baba, S. A. and Hiramoto, Y. (1970) A quantitative analysis of ciliary movement by means of high-speed microcinematography. *J. Exp. Biol.* 52: 675–90.

This analysis examines the hydrodynamics of clam ciliary movement based on precise form changes as determined by high-speed photography. Ciliary waveforms are asymmetrical, and the authors have computed hydrodynamic forces, effectiveness, and bending moments numerically from the data obtained photographically. It is discovered that bend propagation of the cilium occurs in advance of the change in bending moment, thus showing that local active processes are involved in ciliary curvature.

Badenko, L. A., Ivanova, L. V., Kalinin, O. M., Kachurin, A. L., and Kolodyazhnyi, S. F. (1971) A study of the motion of cells of the freshwater sponge *Ephydatia fluviatilis* in the aggregation process. II. Parameters of motion. *Sov. J. Develop. Biol.* 2: 339–43.

This report describes mathematical methods for analyzing parameters of cell motion to yield information on nonrandomness. Both preferential directions of movement and synchronization between adjacent cell aggregates were detected in sponge cell suspensions.

Bailey, N. T. J. (1967) *The Mathematical Approach to Biology and Medicine*, chap. 9, pp. 182–209. New York: Wiley.
   and
Barlett, M. S. (1960) *Stochastic Population Models in Ecology and Epidemiology*, chap. 8, pp. 182–209. London: Methuen.
These works present mathematical treatments and computer-simulated models of the spread of infections, a process in which cell movement or transport is one of the parameters. Despite the complexity of epidemics and the numerous factors that affect the spread of disease, some simulated and deduced temporal patterns do resemble characteristics of specific epidemics.

Blake, J. (1973) A finite model for ciliated micro-organisms. *J. Biomech.* 6: 133–40.
A simple spherical model for the hydrodynamics of a finite ciliated organism is presented. This is achieved by considering two distinct regions: the cilia sublayer and the exterior flow field. The main purpose of this paper is to obtain expressions for the velocity fields through these regions. Graphs of the velocity field are included for two organisms, *Opalina* and *Paramecium*, and are compared with experimental observations for the case of *Paramecium*. Velocities of propulsion of $100\ \mu$/sec for *Opalina* and $1000\ \mu$/sec for *Paramecium* are obtained. Maximum fluid velocities are found at the top of the cilia sublayer.

Blumenson, L. E. (1970) Random walk and the spread of cancer. *J. Theor. Biol.* 27: 273–90.
A mathematic model is constructed to describe the local movements of cancer cells. The application of this theory to clinical observations is reported in Blumenson, Bross, and Slack (1971).[1] Metastasis is assumed to occur by random walk movements through the interstitial spaces as well as by short disseminations through blood vessels. Random walk movements are treated with diffusion kinetics that employ a surrogate diffusion coefficient that can be determined experimentally. The model also accounts for tumor cell proliferation and death; cancer simulations are made of cell spread using the model. Several clinically important aspects are pursued, notably the way in which the model can be used to predict the perimeter of cell metastasis from clinical data.

1. Blumenson, L. E., Bross, I. D. J., and Slack, N. H. (1971) Application of a mathematical model to a clinical study of the local spread of endometrial cancer. *Cancer* 28: 735–44.

Blumenson, L. E., Bross, I. D. J., and Slack, N. H. (1971) Application of a methematical model to a clinical study of the local spread of endometrial cancer. *Cancer* 28: 735–44.
This report describes the clinical usefulness of a mathematical model for spread of cancer cells developed by Blumenson and Bross (*Cancer*, 1971). First it is shown that the model is consistent with clinical data. By using information regarding relative frequency of tumor dispositions both before and after radiation treatment, it was possible to estimate cellular motility and vascular dissemination rates. The deduced interstitial cell motility rates of $60\ \mu$/day compare favorably with malignant cell displacement rates determined in model systems. Finally, the practical usefulness of the model was suggested by analysis of the effectiveness of radiation treatment and by prediction of the clinical effectiveness of early detection.

Brokaw, C. J. (1958) Chemotaxis of brachen spermatozoids. Implications of electro-chemical orientation. *J. Exp. Biol.* 35: 197–212.

Brachen fern spermatozoa exhibit chemotaxis to positive concentration gradients of malate ions. In the presence of malate, and not in its absence, the spermatozoa will also swim toward the anode in an electrical field. Directed swimming is due to the cell changing its direction of swimming under an appropriate influence. Turning rates are proportional to the voltage component perpendicular to the direction of swimming as well as to the perpendicular component of the gradient in the logarithm of malate concentration. Probably both malate and electrical orientations involve the same sensory site. The author suggests that chemotaxis could be explained by localized sites on the fronts of the cells, where malate would reversibly absorb. Interfacial energies will cause forces directing such a site into a concentration gradient, and an electric field would orient this end of a cell toward the anode.

Brokaw, C. J. (1966) Bend propagation along flagella. *Nature* 209: 161–63.

A model for bend propagation in flagella is presented that contrasts with Machin's (1958). In this proposal, bending in one region is recognized to affect neighboring regions because of the finite shear resistance of the flagellum. Bending regions cannot have sharp boundaries. If the passive peripheral bending stimulates active bending forces in this peripheral region, the zone of bend will be propagated along the flagellum.

Brokaw, C. J. (1972a) Flagellar movement: A sliding filament model. *Science* 178: 455–62.

This article presents a lucid account of the historical development of models for flagellar movement, particularly of the way in which ideas have been repeatedly superseded. The author outlines the present form of the sliding filament model. Attention is given to the necessity for, and means of, generating curvature and applying force one-quarter wavelength out of phase with curvature and to the importance in considering load in estimating flagellar activity. The sliding filament model is susceptible to computer simulation, which provides excellent correlation between theoretical and observed flagellar forms. Consideration is also given to mechanochemical processes underlying flagella reaction.

Brokaw, C. J. (1972b) Computer simulation of flagellar movement. I. Demonstration of stable bend propagation and bend initiation by the sliding filament model. *Biophys. J.* 12: 564–86.

A program was developed for digital computer simulation of the movement of a flagellar model consisting of straight segments connected by joints at which bending occurs. This model is presented in Brokaw (1971).[1] The program finds values for the rate of bending at each joint by solving equations that balance active, viscous, and elastic bending moments at each joint. These bending rates are then used to compute the next position of the model. Stable swimming movements, similar to real flagellar movements, can be generated routinely with a 25-segment model using 16 time steps/beat cycle. These results depend on four assumptions about internal flagellar mechanisms: (a) bending is generated by a sliding filament process; (b) the active process is controlled

1. Brokaw, C. J. (1971) Bend propagation by a sliding filament model for flagella. *J. Exp. Biol.* 55: 289–304.

locally by the curvature of the flagellum; (c) nonlinear elastic resistances stabilize the amplitude of the movement; and (d) internal viscous resistances stabilize the wavelength of the movement and explain the relatively low sensitivity of flagellar movement to changes in external viscosity.

Buchsbaum, R., Rashevsky, N., and Stanton, H. E. (1944) A note on the mathematical biophysics of amoeboid movements. *Bull. Math. Biophys.* 6: 61–63.

This note attempts to show that even irregular cell movements, such as those of an amoeba, may be measured and quantitated along simple parameters. It considers the possibility that amoeboid movement may be analogous to periodic elongations under the influence of external forces.

Campbell, R. D. (1967) Tissue dynamics of steady state growth in *Hydra littoralis*. II. Patterns of tissue movement. *J. Morphol.* 121: 19–28.

Epithelial cells are continually undergoing displacement along the column of the *Hydra* polyp as determined by vitally staining some cells and recording their positions on subsequent days. There is some uncertainty as to the pattern of cell proliferation that is the source of the displacing cells. In this analysis an expression is derived relating the distribution of cell proliferation to the pattern of displacement. With information about the cell cycle time, but not without, the pattern of cell movement provides a reasonably sensitive measure of proliferation pattern.

Carlson, F. D. (1959) The motile power of a swimming spermatozoan. In *Proceedings of the First National Biophysics Conference, Columbus, Ohio, 1957*, ed. H. Quastler and H. J. Morowitz, pp. 443–49. New Haven: Yale University Press.

The energy exerted to move the tail and that exerted to move the head of a sea urchin spermatozoan are calculated separately and found to add up to 3×10^{-7} erg/sec, about one-quarter of the metabolic energy thought to be consumed.

Carter, S. B. (1967) Haptotaxis and the mechanism of cell motility. *Nature* 213: 256–60.

This interesting paper presents and discusses a theory of cell motility based on interfacial tensions. The mathematical basis is treated briefly by Moilliet (1967).[1] The theory holds that the edge of a cell in contact with a substratum will passively (in the biological sense) advance or retract depending on the interfacial tensions of the three surfaces that meet there: substrate–cell, substrate–medium, and cell–medium. Carter further postulates that the cell modifies the substratum that it contacts by depositing a microexudate; this layer decreases the substrate–medium interfacial tension and hence promotes cell retraction on all edges except the leading one. This leading margin is in advance of the microexudate and hence edge advancement is promoted. Evidence for this theory is provided by the phenomenon of haptotaxis; when cells are placed on a substratum of graded adhesivity, they move up the gradient. Carter has presented an analysis of the "ruffled membrane" that is quite different from the one currently held by cytologists.

1. Moilliet, J. L. (1967) Elementary surface thermodynamics of Carter's theory of haptotactic cell movement. *Nature* 213: 260–61.

Chwang, A. T., Wu, T. Y., and Winet, H. (1972) Locomotion of *Spirilla*. *Biophys. J.* 12 : 1549–61.

The hydromechanics of *Spirillum* locomotion is analyzed by considering the balance of both rectilinear and angular momenta of the surrounding viscous fluid, which is otherwise at rest. The physical model of *Spirillum* adopted for the present analysis consists of a rigid helical body with flagella attached to both ends of the helix. The motion is supposed to be activated first by the polar flagella, both rotating in the same sense, thus causing the helical body to rotate in the opposite sense in angular recoil, which in turn pushes the body forward in response to the balance of linear momentum of the surrounding fluid. The sweeping back of the polar flagella during forward motion is ascribed to a certain bending flexibility of the flagella and to their conjunction with the body. Based on this model, some quantitative results for *Spirillum* movement are predicted and are found to be consistent with existing experimental data.

Coakley, C. J. and Holwill, M. E. J. (1972) Propulsion of micro-organisms by three-dimensional flagellar waves. *J. Theor. Biol.* 35 : 525–42.

This paper analyzes the hydrodynamics of flagella with nonuniform, three-dimensional waveforms. Significant errors arise in estimations of power dissipation when one uses average waveform values in considering, for example, flagella that show progressive alteration in λ from base to tip. This research establishes equations and representative numerical solutions for efficiency, velocity, and power expenditure of flagellar waveform families as functions of ellipticity and nonuniformity. The authors also make a significant examination of rotational couple at the flagellar–head junction and discuss the energy dissipation associated with this event.

Cohen, D. and Eilam, G. (1970) Computer simulation of biological pattern generation by purely local interactions: lobed and smooth boundaries. *Comput. Biol. Med.* 1 : 117–23.

The authors test the ability of local cell motility rules to produce uniform global patterns. In this analysis, which deals with rings of cells, it was found using computer simulation that simple local motility rules can lead to the formation of circles or lobed patterns. The motility rules employed were based only on a cell's position relative to its two closest neighbors. Thus, large-scale biological patterns do not imply the existence of information flowing across an entire structure.

Cohen, M. H. and Robertson, A. (1971a) Wave propagation in the early stages of aggregation of cellular slime molds. *J. Theor. Biol.* 31 : 101–18.

A detailed theory of the propagation velocity of the acrasin pulse responsible for the aggregation of some of the cellular slime molds is presented. The ingredients are diffusion of acrasin, triggering after a threshold concentration is reached, an intracellular delay between the reaching of threshold and subsequent release of acrasin, the destruction of acrasin by acrasinase, and an intracellular refractory period. It is suggested that the rate-limiting factor in the velocity of propagation is probably the intracellular delay and not intercellular diffusion. It is shown that there is a critical density of amoebas below which the waves cannot propagate. The number of neighboring amoebas triggered simultaneously by the propagating wave, and therefore the spatial width of the pulse, tends to increase with density. Rough estimates are given for the number of acrasin molecules per pulse and for its threshold concentration.

Cohen, M. H. and Robertson, A. (1971b) Chemotaxis and the early stages of aggregation in cellular slime molds. *J. Theor. Biol.* 31: 119–30.

The authors consider the effect of aggregative movement on organizing wave propagation in the cellular slime molds. Chemotactic responsiveness is formulated mathematically, and quantitative results from observations on slime mold cells are correlated with the theoretical analysis. The chemotactic signal is impulsive, that is, short compared to the duration of the response. The response is of the all-or-none variety. Differences in period between *Dictyostelium discoideum* and *Polysphondylium violaceum* are compensated by differences of chemotactic response to give comparable aggregation territories. Consideration is also given to wave propagation and cell density changes produced by chemical response to the periodic signals.

Curtis, A. S. G. (1970) Problems and some solutions in the study of cellular aggregation. *Symp. Zool. Soc. London* 25: 335–52.

An expression for heterotypic collision frequency in mixtures of cell suspensions is presented. The expression is a function of the fluid shear rate in the environment of the cells, a process that can be controlled sufficiently carefully by means of a Couette viscometer. The author uses this model in an attempt to devise a new way of measuring cellular adhesiveness. This analysis is also discussed in Curtis (1969).[1]

1. Curtis, A. S. G. (1969) The measurement of cell adhesiveness by an absolute method. *J. Embryol. Exp. Morphol.* 22: 305–25.

Dahlquist, F. W., Lovely, P., and Koshland, D. E. (1972) Quantitative analysis of bacterial migration in chemotaxis. *Nature New Biol.* 236: 120–23.

Important experimental approaches are devised and described that permit quantitative analysis of bacterial chemotactic movement. The apparatus consists of a chamber along which bacterial and chemoattractant concentrations can be established in predetermined patterns; the subsequent redistribution of bacteria is sensed by light scattering and is recorded. This report establishes that bacterial migration velocities are approximately proportional to the relative concentration gradient $(dc/c) dx$ of the attractant substance. There is some dependence upon the absolute concentration, however. Bacterial chemotactic migration velocities (1–3 μ/sec) are about $\frac{1}{10}$ the velocity of bacterial swimming, so that chemotaxis may involve a biased random walk, as suggested by Armstrong, Adler, and Dahl (1967)[2] and Keller and Segal (1971).[3] The authors believe that this experimental system provides an analog for studying the kinetics of sensory physiology. A parallel experimental and theoretical program is described by Nossal and Chen (1973).[4]

2. Armstrong, J. B., Adler, J., and Dahl, M. A. (1967) Nonchemotactic mutants of *Escherichia coli*. *J. Bacteriol.* 93: 390–98.
3. Keller, E. F. and Segel, L. A. (1971) Model for chemotaxis. *J. Theor. Biol.* 30: 225–34.
4. Nossal, R. and Chen, S. H. (1973) Effects of chemo-attractants on the motility of *Escherichia coli*. *Nature New Biol.* 244: 253–54.

Doerner, K. (1967) Pressure gradients in cells. *J. Theor. Biol.* 14: 284–92.

Fundamental stress formulas for the thick-walled sphere apply to the spherical cell. The cytoplasm is the thick wall and the nucleus is the core. The modulus of elasticity in the sphere must be constant for use of these formulas. The heterogeneous cytoplasm can be converted mathematically to a material with constant modulus. This conversion is along a radius and does not include the surface variations on each differential shell. When the cytoplasm converts to a constant material, the stress formulas show that a significant internal fluid pressure exists in the cytoplasm. Depending on the phase condition of the cell, this internal pressure produces tension, compression, or a mixture of stresses in the cytoplasm and cellular membrane. Imbalances of these types may be related to cellular deformations during movement.

Ede, D. A. and Law, J. T. (1969) Computer simulation of vertebrate limb morphogenesis. *Nature* 221: 244–48.

Cell division and movement in chick-limb development were simulated by computer reiteration of simple mitosis and movement rules. Motility rules involved a linear biasing of the location in which a newly divided cell was placed. Gradients in proliferative rates across the tissue were also simulated. The authors conclude that normal limb development can be successfully modeled in this way and that slight changes in motility parameters lead to results reminiscent of limb development in a mutant chicken strain, *talpid*[3].

Fraenkel, G. W. and Gunn, D. L. (1961) *The Orientation of Animals: Kineses, Taxes, and Compass Reactions.* Dover: New York.

This book deals extensively with patterns of locomotion, mainly as exhibited by multicellular organisms. One example of a model describing motility of single cells is the "triangle of forces" equation, which is applied to the motile alga *Euglena* (chap. 11). Suppose that an organism has a flat eyespot on each side and that its motion in the vicinity of two light sources is such as to maintain equal the illumination falling on the two eyespots. The interesting pathway described by such an organism's locomotion is closely simulated by *Euglena*'s motion. Although *Euglena* has but a single eyespot, the organ is displaced to one side and the animal's constant rotation may render the eye equivalent to a bilateral sensory device.

Gail, M. H. and Boone, C. W. (1970) The locomotion of mouse fibroblasts in tissue culture. *Biophys. J.* 10:980–93.

This paper reports an experimental and theoretical investigation into the randomness of cell locomotion. It is established that fibroblasts tend to persist in their direction of motion from one 2.5-hour time interval to the next. Over successive 5-hour intervals, however, persistence was no longer detectable. A theoretical investigation of persisting cells revealed that their motility can be characterized by a single constant, D^*, the augmented diffusion constant. A parameter model was constructed that is in excellent agreement with observed mean square displacements. Methods for experimentally determining D^* and for establishing confident limits for this constant were also developed. A random walk model, modified to include the persistence effect, was thus found to describe the motion of fibroblasts in culture. The use of this model in quantitating decrease in motility by susceptible cell strains at high density is presented in Gail and Boone (1971).[1]

1. Gail, M. H. and Boone, C. W. (1971) Density inhibition of motility in 3T3 fibroblasts and their SV40 transformants. *Exp. Cell Res.* 64: 156–62.

Gittleson, S. M. and Jahn, T. L. (1968) Vertical aggregations of *Polytomella agilis*. *Exp. Cell Res.* 51 : 579–86.

The flagellate protozoan *Polytomella agilis* forms dense populations at the surface of the growth medium. Teardrop-shaped aggregates irregularly sink down from the surface at rates (almost 1 mm/sec) several orders of magnitude greater than the sinking rate for an individual cell. Apparently the drops of cells fall because their specific gravity, due to extreme accumulation of cells, becomes significantly greater than that of the underlying culture medium. In agreement with this phenomenon, the falling kinetics fit Stokes's law fairly well, with the use of experimentally measured values for cell specific gravity and population density. Patterned aggregates of this sort were extensively studied qualitatively by Wager (1911).[1]

1. Wager, H. (1911) On the effect of gravity upon the movements and aggregation of *Euglena viridis* Ehrb., and other micro-organisms. *Phil. Trans. Roy. Soc. London B* 201 : 333–90.

Goel, N., Campbell, R. D., Gordon, R., Rosen, R., Martinez, H., and Yčas, M. (1970) Self-sorting of isotropic cells. *J. Theor. Biol.* 28 : 423–68.[2]

Some conditions and consequences of Steinberg's (1963) hypothesis[3] on the self-sorting of cells are examined. The system used is modeled as a two-dimensional grid whose squares represent cells or ambient medium. Adhesivity is simulated by the strength λ of each cell–cell contact, as well as by the total energy function (E) for the grid. All the configurations (patterns) with maximum E have been found for two- and three-cell types, together with the constraints on the values of various λs. None of these configurations is histologically interesting. Therefore the concepts of neighboring configurations and patterns representing local E maxima were explored; some of the emergent patterns are histologically interesting. Finally, computer simulation of cell sorting out shows the presence of many trapping configurations representing local E maxima. Therefore Steinberg's hypothesis requires special rules of cell motility, which remain to be determined.

2. Paper 7 in this book.
3. Steinberg, M. S. (1963) Reconstruction of tissues by dissociated cells. *Science* 141 : 401–08.

Goel, N. S. and Leith, A. G. (1970) Self-sorting of anisotropic cells. *J. Theor. Biol.* 28 : 469–82.[4]

This is a continuation of the Goel et al. (1970) investigation[5] of the implications of Steinberg's (1963) differential adhesion model of cell sorting out.[6] In this analysis cells are considered to be anisotropic—that is, different margins have different adhesive qualities. Cells are still considered nondeformable members of a tessellation. It is shown that some patterns representing maximal E (energy of adhesion) values mimic histologically interesting structures, for example, tubules, sheets, and vesicles.

4. Paper 8 in this volume.
5. Goel, N. S. et al. (1970). Self-sorting of isotropic cells. *J. Theor. Biol.* 28 : 423–68.
6. Steinberg, M. S. (1963) Reconstruction of tissues by dissociated cells. *Science* 141 : 401–08.

Good, R. J. (1972) Theory of the adhesion of cells and the spontaneous sorting-out of mixed cell aggregates. *J. Theor. Biol.* 37 : 413–34.

The physicochemical basis of cell adhesion is examined with regard to the mechanisms of cell sorting out. Of the several parameters that will affect relative adhesive energies between two cells, the surface potentials will be the most influential in biological circumstances. It is shown that the replacement of heterotypic by homotypic bonds (i.e., apposition of surfaces) can be thermodynamically favorable and that these considerations *could* provide a nonspecific basis for cell sorting. The article also presents a critical appraisal of current physical theories of cell adhesion.

Gordon, R. and Drum, R. W. (1970) A capillarity mechanism for diatom gliding locomotion. *Proc. Nat. Acad. Sci. U.S.* 67: 338–44.

To explain diatom locomotion, these authors examine a mechanochemical process (postulated by Drum and Hopkins, 1966)[1] that has not previously been applied to cellular movement. Diatom shells are perforated by slits called raphae. It is proposed that the diatom raphe is a parallel-plate capillary containing a fluid that reacts at the trailing end, turning into a form that no longer wets the raphe walls and that is left behind as a trail. More unreacted raphe fluid is drawn by capillary pressure from a source near the leading end of the raphe. This fluid sticks out from the raphe along its length, adhering to surfaces and thus causing gliding locomotion. Formulas are given for the maximum velocity and force of a moving diatom as a function of raphe dimensions and the surface tension and velocity of the fluid. An *a priori* estimate of the force exerted by a moving diatom, 1–50 mdyne, agrees with measured values.

1. Drum, R. W. and Hopkins, J. T. (1966) Diatom locomotion: an explanation. *Protoplasma* 62: 1–32.

Gordon, R., Goel, N. S., Steinberg, M. S., and Wiseman, L. L. (1972) A rheological mechanism sufficient to explain the kinetics of cell sorting. *J. Theor. Biol.* 37: 43–73.[2]

Most previous efforts to explain cell sorting in mixed reaggregates have relied on the assumption that cell motility is necessary. In this analysis the suggestion is made that only minute movements may be required, so small in fact that they should be considered as changes in cell shape rather than cell locomotion. The driving forces for these shape changes are local increases in intercellular adhesion promoted by the changes. The changes are resisted by the viscosity of the cell aggregate. This viscosity, estimated several ways, is extraordinarily high: on the order of 10^7 poise. The authors conclude that the viscosity reflects slippage between rough cell membranes and discuss the implications of such a high viscosity on the kinetics of cell sorting out.

2. Paper 11 in this book.

Gray, J. (1953) Undulatory propulsion. *Quart. J. Microscop. Sci.* 94: 551–78.

This paper, while not specifically modeling any particular cell's motility, sets out the basic considerations that relate to undulatory propulsion. A particularly relevant section deals with the nature of helical waves and the characteristics they must satisfy in order to result in propulsion. This article deals with both macroscopic and microscopic terrestrial and aquatic movement.

Gray, J. (1958) The movement of the spermatozoa of the bull. *J. Exp. Biol.* 35:96–108.

The detailed waveforms and geometry of bull spermatozoa are analyzed with the aid of high-speed photography. These spermatozoa do not have the simplicity of motion exhibited by sea-urchin spermatozoa.[1] Rather, the wavelength and amplitude change continuously as the wave propagates down the tail, and the cycle includes some forward movement in parts of the flagellum. The methods of Gray and Hancock (1955)[2] are used to calculate several parameters of the propulsive forces generated by different flagellar regions, but it was considered impracticable to relate the speed of propulsion to the complex waveforms in this species.

1. As detailed in Gray, J. (1955) The movement of sea-urchin spermatozoa. *J. Exp. Biol.* 32: 775–801.
2. Gray, J. and Hancock, G. J. (1955) The Propulsion of Sea-Urchin Spermatozoa. *J. Exp. Biol.* 32: 802–14.

Gray, J. and Hancock, G. J. (1955) The propulsion of sea-urchin spermatozoa. *J. Exp. Biol.* 32: 802–14.
The general theory of flagellar propulsion is discussed and an expression obtained whereby the propulsive speed of a spermatozoan can be expressed in terms of the amplitude, wavelength, and frequency of the waves passing down the tail of a spermatozoan of the sea urchin, *Psammechinus miliaris*. The expression obtained is applicable to waves of relatively large amplitude and assumes that waveforms are sinusoidal. The calculated propulsive speed is almost identical with that derived from observational data. Unless the head of a spermatozoan is very much larger than that of *Psammechinus*, its presence makes relatively little difference to the propulsive speed. Most of the energy of the cell is used up in overcoming the tangential drag of the tail. Although the amplitude may change as a wave passes along the tail, the propulsive properties of the latter may be expected to be closely similar to those of a tail generating waves of the same average amplitude. This analysis has served as the starting point for most subsequent studies of flagellar movement.

Hancock, G. J. (1953) The self-propulsion of microscopic organisms through liquids. *Proc. Roy. Soc. London A* 217: 96–121.
A new model is presented for analyzing the flagellar propulsion that overcomes a limitation of Taylor's model,[3] which is valid only for lower amplitude displacements than are observed in biology. The new model is based on Stokes's solution for flow past a sphere and is valid for large amplitudes, subject to the condition that the waving filament has a diameter small compared to the wavelength. Expressions are obtained for velocity of propagation of filaments with planar and helical waveforms. The predictions of this model and that of Taylor[3] are in agreement under the conditions where their validities overlap.

3. Taylor, Sir G. (1952) The action of waving cylindrical tails in propelling microscopic organisms. *Proc. Roy. Soc. London A* 211: 225–39.

Holtfreter, J. (1943) A study of the mechanics of gastrulation. *J. Exp. Biol.* 94: 261–318.
This pioneering study combines observations on dissociated and reassociated embryonic tissue with perceptive analysis of possible mechanisms underlying the tissue movements and specificity. The concept of spreading films as analogs of spreading tissues is discussed, and it is suggested that interfacial tensions are responsible for the enveloping behavior of certain tissues, the nonspreading observed in other tissue combinations, and a variety of other tissue activities such as blastomere positioning and tissue rejection.

This study has apparently led directly to Steinberg's (1963) differential adhesion hypothesis of cell sorting.[1]

1. Steinberg, M. S. (1963) Reconstruction of tissues by dissociated cells. *Science* 141 : 401–08.

Holwill, M. E. J. (1965) The motion of *Strigomonas oncopelti*. *J. Exp. Biol.* 42 : 125–37.

The movement of the flagellum of *Strigomonas oncopelti* was studied using high-speed cinephotography. Waves usually pass along the flagellum from tip to base, but under certain conditions waves are propagated from base to tip, reversing the direction of movement of the organism. Increasing the viscosity of the medium reduces the frequency of beat, while the shape of the wave remains unaltered. The variations in the wave parameters with increasing viscosity are consistent with a mechanically propagated wave. Calculated values for Young's modulus indicate that the membrane, the matrix, the peripheral fibrils, or the central pair could act as the compressive element needed to resist bending. The mechanism controlling the direction of wave propagation appears to lie in the membrane.

Holwill, M. E. J. (1966) Physical aspects of flagellar movement. *Physiol. Rev.* 46 : 696–785.

In this valuable review, the hydrodynamic models of flagellar motion are set forth and correlated with the structure of flagella. Both eukaryotic and bacterial organs are treated. There is also a valuable discussion of thermodynamic considerations of flagellar activity.

Holwill, M. E. J. and Burge, R. E. (1963) A hydrodynamic study of the motility of flagellated bacteria. *Arch. Biochem. Biophys.* 101 : 249–60.

Flagellated bacteria have been considered to propel themselves either by active helical movements of the flagella or by active movements of the hellically shaped body. These two competing hypotheses are examined hydrodynamically in this paper. The discussion hinges on the question of whether flagella are the cause or the effect of motility. Expressions are derived for the velocities of propulsion of a bacterium in the two cases, following the methods introduced by Gray and Hancock (1955).[2] The expenditure of energy necessary to maintain helical waves of displacement is compared with the energy associated with sinusoidal waves. The results show clearly that the magnitude of the observed velocities (up to 50 μ/sec) can only be explained if the flagella are active motor organs. Only a few percent of the flagellar energy is employed in driving the body through the medium; the balance is expended to maintain the helical motion of the flagella.

2. Gray, J. and Hancock, G. J. (1955) The propulsion of sea-urchin spermatozoa. *J. Exp. Biol.* 32 : 802–14.

Holwill, M. E. J. and Miles, C. A. (1971) Hydrodynamic analysis of non-uniform flagellar undulations. *J. Theor. Biol.* 31 : 25–42.

The velocity of translation and power expenditure against viscous resistance of nonuniformly beating flagella are calculated. If the mean wave parameters of flagella bearing waves of varying amplitude and wavelength are used, reasonable approximations to the limiting propulsive speed can be obtained by using equations developed by earlier workers. However, the power expenditure against viscous resistance may be significantly greater than that calculated on the basis of the mean wave parameter assumption.

Keller, E. F. and Segel, L. E. (1970) Initiation of slime mold aggregation viewed as an instability. *J. Theor. Biol.* 26:399–415.

These authors have defined a model of cellular slime mold aggregation that assumes simple kinetics for the following parameters: acrasin (chemoattractant) production, diffusion, and destruction; acrasinase turnover; and cell motility and response to acrasin. The mathematical formulation of the interaction of these several processes is shown to have stable homogeneous solutions. Perturbations can result in instabilities that imply cell aggregation. Oscillatory behavior and conditions promoting an altered cell physiology are also predictable from this model, in agreement with experimental observation.

Keller, E. F. and Segel, L. E. (1971) Model for chemotaxis. *J. Theor. Biol.* 30: 225–34.

The chemotactic response of unicellular microscopic organisms is viewed as analogous to Brownian motion. Local assessments of chemical concentrations made by individual cells give rise to fluctuations in path. When averaged over many cells or a long time interval, a macroscopic flux proportional to the chemical gradient is derived. By way of illustration, the coefficients appearing in the macroscopic flux equations are calculated for a particular microscopic model.

Leith, A. G. and Goel, N. S. (1971) Simulation of movement of cells during self-sorting. *J. Theor. Biol.* 33: 171–88.[1]

This paper investigates the effects of various cell motility rules on the ability of cells to sort out in mixed aggregates, according to the differential adhesion model of Steinberg (1963).[2] The analysis uses the general methodology established by Goel et al. (1970)[3] of a tessellation analogy of the cell population and involves computer simulation to predict the resulting patterns of cell sorting. Allowing cells to move with some independence of the local adhesive energy can lead to more complete sorting out than in previous models,[3] and quantitative relations are presented between the degree of sorting out and the motility characteristics of the cells.

1. Paper 9 in this book.
2. Steinberg, M. S. (1963) Reconstruction of tissues by dissociated cells. *Science* 141 : 401–08.
3. Goel, N. S. et al. (1970) Self-sorting of isotropic cells. *J. Theor. Biol.* 28 : 423–68.

Leontovich, A. M., Pyatetskii-Shapiro, I. I., and Stavskaya, O. N. (1970) Certain mathematical problems related to morphogenesis. *Avtomatika i Telemekhanika* 4: 94–107.[4]

This work initiates a general examination of morphogenesis based on local, homogeneous cell motility rules. The authors center this analysis on the problem of straightening (linearizing) a curved file of cells. Cellular motility rules are based solely on a cell's position relative to its nearest neighbor on each side. The general formulation of motility rules is established and the generality of possible rules is examined. Using computer simulation, the linearization is examined as a function of starting configuration.

4. Paper 1 in this volume.

Leontovich, A. M., Pyatetskii-Shapiro, I. I., and Stravskaya, O. N. (1971) The problem of circularization in mathematical modeling of morphogenesis. *Avtomatika i Telemekhanika* 2: 100–10.[1]

In this paper, a sequel to Leontovich et al. (1970),[2] the mathematics and kinetics of circularization are investigated. Circularization is considered as the transformation of *n* points into apices of a regular polygon. As in their previous paper, the authors deal only with local, homogeneous motility rules that are iterated in discrete time. General expressions for several classes of motility functions are established and explored. Using computer simulation, the investigators examined the effectiveness of various motility rules, the rates at which circularization proceeded with different motility functions, and the dependence of the process on the initial configuration of the points.

1. Paper 2 in this volume.
2. Leontovich, A. M., Pyatetskii-Shapiro, I. I., and Stavskaya, O. N. (1970) Certain mathematical problems related to morphogenesis. *Avtomatika i Telemekhanika* 4: 94–107.

Lew, H. S. and Fung, Y. C. (1970) Plug effect of erythrocytes in capillary blood vessels. *Biophys. J.* 10: 80–99.

As an idealized problem of the motion of blood in small capillary blood vessels, the low Reynolds number flow of plasma in a circular cylindrical tube involving a series of circular disks is studied. It is assumed for this study that the suspended disks are equally spaced along the axis of the tube and that their faces are perpendicular to the tube axis and are centered. The inertial force of the fluid due to convective acceleration is neglected on the basis of the smallness of the Reynolds number. The solution of the problem is derived for a quasisteady flow involving infinitesimally thin disks. The numerical calculation is carried out for a set of different combinations of the interdisk distance and the ratio of the disk radius to the tube radius. The ratio of the velocity of the disk to the average velocity of the fluid is calculated. The plasma viscosity, pressure gradients, and different rates of transport of red blood cells and of plasma in the capillary blood vessels are computed and discussed.

Lewis, J. (1973) The theory of clonal mixing during growth. *J. Theor. Biol.* 39: 47–54.

Lewis, J. H., Summerbell, D., and Wolpert, L. (1972) Chimeras and cell lineage in development. *Nature* 239: 276–79.

These papers explore the term *tissue progenitor cell* and the adequacy of previous attempts to determine the number of progenitor cells by means of genetic mosaics. The term progenitor cells may be defined to include a variety of cells, ranging from those that first contribute some descendants to a tissue (the zygote itself could thus be a progenitor) to those that are present when the tissue is essentially completed and differentiated. There are intermediate definitions—those including cells whose descendants are only partly included in the tissue. The number of separate cell clones that will be present in a mosaic individual depends not only on the definition of progenitor, but also on the extent of cell migration and on the relative sizes of the cell clones and the tissue rudiment. The variability of clone number observed in corresponding tissues of similar animals will also depend on cell movements and clone sizes. The kinetics of the growth of clonal boundaries and mixing of the cells are examined mathematically.

Previous experimental analyses of progenitor cell numbers have not adequately recognized these complexities.

This analysis casts doubt on the conclusion, drawn from experiments,[1] that vertebrate embryonic structures have fixed numbers of precursor cells. It is shown here, and implicitly suggested, that the apparent determinate primordial cell number could represent instead the average number of discrete cell clones existing at the time that clonal mixing becomes ineffective.

1. Mintz, B. (1967) Gene control of mammalian pigmentary differentiation. I. Clonal origin of melanocytes. *Proc. Nat. Acad. Sci. U.S.* 58: 344–51.

Lotka, A. J. (1956) *Elements of Mathematical Biology*, reprint, 465 pp. New York: Dover.

This rewarding book touches on many aspects of modeling. One essay (pp. 358 ff) applicable to cellular movement consists of a statistical mechanical view of organism movement, migration, collision frequency, and persuit. The author refers the reader to Bachelier's *Calcul des Probabilités* for a discussion of basic random walk motions.

Lubliner, J. and Blum, J. J. (1971) Model for bend propagation in flagella. *J. Theor. Biol.* 31: 1–24.

Some previous workers have suggested that active bending processes in flagella are activated by passively propagated bending. This idea is incorporated into a theoretical model that describes the effect of viscosity on the shape and propagation velocity of flagellar bends. The model incorporates a first-order active bending process that is initiated when a critical level of passive bending is reached and a first-order unbending process that follows after a prescribed time interval. The effects of the external viscosity and the internal mechanical properties of the flagellum are included in such a way that it is possible to predict the velocity of bend propagation and the shapes of the bending and unbending transitions for steady waves on an infinite flagellum. These predictions are compared with published data on the effect of viscosity on the velocity of bend propagation along flagella, and values for all the parameters of the flagellar model have been estimated.

Lubliner, J. and Blum, J. J. (1972) Model of flagellar waves. *J. Theor. Biol.* 34: 515–34.

This paper extends the authors' model (Lubliner and Blum 1971)[2] of flagellar bend propagation. This form of the model allows active contraction to occur in both top and bottom fibers in periodic alternation. The extension permits computation of the complete wave shape from first principles. Calculated wave parameters are compared with the experimental measurements of Brokaw (1966).[3] Good agreement is obtained for living sperm of the sea urchin *Lytechinus* and for glycerinated sperm of *Lytechinus* reactivated in 2.0 μM ATP. A range of agreement was found between calculations based on this model and experimental observations on sperm of two other animals.

2. Lubliner, L. and Blum, J. J. (1971) Model for bend propagation in flagella. *J. Theor. Biol.* 31: 1–24.
3. Brokaw, C. J. (1966) Effects of increased viscosity on the movements of some invertebrate spermatozoa, *J. Exp. Biol.* 45: 113–39.

Machin, K. E. (1958) Wave propagation along flagella. *J. Exp. Biol.* 35: 796–806.

This article considers the flagellar waveform parameters implied by two separate models of flagellar bending: (1) that oscillatory energy is supplied only at the proximal end, which is rigidly fixed to the body, and (2) that active bending occurs and is propagated throughout the length of the flagellum. This analysis deals only with planar waveforms and for waveforms produced under conditions of low Reynolds number. It is found that no known naturally occurring waveforms can be produced by the proximally hinged model; in this model, for example, waveforms always damp out in less than $1\frac{1}{2}$ wavelengths. The homogeneous model produces waveforms similar to those observed experimentally. This implies that contractile processes occur along the length of flagella.

Machin, K. E. (1963) The control and synchronization of flagellar movement. *Proc. Roy. Soc. London B* 158: 88–104.

Waves of bending will arise spontaneously on a flagellum if changes in the length of its contractile elements cause changes in tension after a delay. The properties of the contractile elements must be nonlinear if the amplitude of these waves is to remain finite. This nonlinearity has two consequences: (1) control of the frequency and direction of propagation of the waves can be exercised from the proximal end of the flagellum, and (2) two nearby flagella will synchronize, beating with a common frequency and wavelength. Synchronization has been observed among spermatozoa.

Moilliet, J. L. (1967) Elementary surface thermodynamics of Carter's theory of hapto-tactic cell movement. *Nature* 213: 260–61.

This work supports the interfacial thermodynamic theory of Carter.[1] It analyzes the contact angle and displacement energetically favored at an edge where cell, substratum, and medium meet. The most interesting aspect of this report is a brief extension of the thermodynamics to include the case where a surface membrane is present around the cell.

1. Carter, S. B. (1967) Haptotaxis and the mechanism of cell motility. *Nature* 213: 256–260.

Nossal, R. (1972*a*) Boundary movement of chemotactic bacterial populations. *Math. Bio. Sci.* 13: 397–406.

Chemotactic bacteria inoculated locally in a medium will exhibit moving population growth boundaries if the medium contains a chemoattractant that is metabolized (see Adler and Dahl, 1967).[2] In this paper expressions are derived for the kinetics of these growth boundaries under simple conditions. It is assumed that chemotactic movements are proportional to the logarithm of metabolite concentration gradients. A subsequent paper (Nossal 1972*b*)[3] provides experimental data on bacterial movement in agar that verify the applicability of this model.

2. Adler, J. and Dahl, M. M. (1967) A method for measuring the motility of bacteria and for comparing random and non-random motility. *J. Gen. Microbiol.* 46: 161–73.
3. Nossal, R. (1972*b*) Growth and movement of growth rings of chemotactic bacteria. *Exp. Cell Res.* 75: 138–42.

Peterson, S. C. and Noble, P. B. (1972) A two-dimensional random-walk analysis of human granulocyte movement. *Biophys. J.* 12: 1048–55.

This paper derives an expression for distance traversed by a random walk of non-uniform step lengths. This formulation is designed as a method for determining randomness of cell movement using discrete time observations as with cinemicroscopy. The method assesses randomness without measurements of directionality and discriminates between granulocyte motion in the absence and presence of bacteria that invoke chemotactic motion.

Phillips, H. M. (1969) "Equilibrium Measurements of Embryonic Cell Adhesiveness; Physical Foundation and Testing of the Differential Adhesion Hypothesis." Ph.D. thesis, Johns Hopkins University, Baltimore, Md.

Cell aggregates subjected to prolonged centrifugal force alter in shape because of cell movement. The equilibrium configuration is reached when the adhesive forces between the cells, tending to minimize the surface area of the aggregate, balance the centrifugal force, which tends to flatten the aggregate. This elegant study was undertaken to obtain measurements of the cohesive forces between cells.

Poodry, C. A., Bryant, P. J., and Schneiderman, H. A. (1971) The mechanism of pattern reconstruction by dissociated imaginal discs of *Drosophila melanogaster. Develop. Biol.* 26: 464–77.

Segregated regions of future adult tissue (imaginal discs) exist in many insect larvae, and these discs are widely used in studying pattern formation. Discs dissociated into cells or cell clumps can be reaggregated, and the resulting mosaic tissue patterns have prompted some workers to suppose that the reassociating cells migrate extensively through the reaggregating mass. These authors examine the frequency of association between cells (and cell clumps) from discs of two genotypes. They also model the reassociation with a computer simulation of a probabilistic analog. It is concluded that cells probably do not reassociate with other cells outside their 6–8 nearest neighbors.

Rashevsky, N. (1939) Mathematical biophysics of the cell with reference to the contractility of tissues and amoeboid movements. *Bull. Math. Biophys.* 1: 47–62.

This analysis investigates the action of metabolite diffusion forces on cell deformation when the deformation occurs at least as rapidly as the molecular flows. There exist possible configurations with several equilibria as well as with periodical contractions and expansions around one such configuration. These effects are viewed as possible mechanisms in causing cellular movements, such as the formation of pseudopodia, and this analysis provides a general background for several later papers by this author.

Rashevsky, N. (1940a) Physicomathematical aspects of some problems of organic form. *Bull. Math. Biophys.* 2: 109–21.

This essay traces polarity of egg cytoplasm (due, for example, to a graded metabolite) through cleavage and shows that its pattern may become more and more complex. Finally, gastrulation pattern and forces can be derived directly from this polarity via mechanisms of differential growth or patterned surface tensions.

Rashevsky, N. (1940*b*) Contributions to the mathematical biophysics of organic form. III. Deformation of shell-shaped cellular aggregates. *Bull. Math. Biophys.* 2:123–6.
This article, continuing the analysis of the previous paper, shows that the metabolism of a material related to the gradient can give rise to mechanical pressure patterned so as to result in gastrulation of an embryo. The argument is based on regional production and flow of a material through a surface of the embryo that exerts a pressure the author estimates by means of the ideal gas law.

Rashevsky, N. (1940*c*) Deformation of shell-shaped cellular aggregates: application to gastrulation. *Bull. Math. Biophys.* 2: 169–75.
This investigation extends the discussion of gastrulation from the two preceding publications. Here more explicit calculations are made for the flow of materials generating the deformations. A more extensive discussion of the geometry of deformation is also presented.

Rashevsky, N. (1960) Mathematical biophysics. In *Physico-Mathematical Foundations of Biology*, 3rd rev. ed. 2 vols. New York: Dover.
Many topics dealt with in this volume treat modeling of biological processes. Chapter 27 recapitulates the author's ideas on the origins of forces responsible for gastrulation forces and other developmental movements and deformation.

Rikmenspoel, R. (1971) Contractile mechanisms in flagella. *Biophys. J.* 11 : 446–63.
The elastic theory of flexural waves in thin rods accurately predicts the velocity of flagellar bending waves over a wide range of viscosities. This shows that flagella behave as a purely mechanical system for the transmission of these waves. An evaluation of the total bending moment reveals that this moment occurs in phase over the entire length of a flagellum. From this it is concluded that each contractile fiber in the flagella is activated simultaneously over its whole length. The magnitude of the bending moment decreases linearly along the flagellum. This relationship is most easily explained by a sliding filament hypothesis in flagella with the elementary 9 + 2 fibers. The expression found for the bending moment explains logically that the wave velocity in flagella is determined by only their mechanical properties and the outside viscosity.

Roberts, A. M. (1970*a*) Motion of spermatozoa in fluid streams. *Nature* 228:375–76.
Bull spermatozoa exhibit positive rheotaxis (their swimming path is aligned upstream) under certain conditions of fluid flow. The author postulates that this orientation results from gravitational torque, which tends to align the spermatozoa head downward, and from flow-orientation torque, which tends to align them in the direction of the streamlines. Quantitative expressions for these influences are given, and the author notes the possible significance of this behavior in *in vivo* sperm transport and fertilization.

Roberts, A. M. (1970*b*) Geotaxis in motile micro-organisms. *J. Exp. Biol.* 53: 687–99.
This article suggests that the principal cause of geotaxis in many motile microorganisms is a gravity induced hydrodynamic torque, the size of which depends upon the shape of the organism. The nature of the torque was investigated by the use of small scale models. An expression for cell distributions in long vertical columns is derived and compared with experimentally determined values using *Paramecium*. It is shown that the strength of the hydrodynamic interaction is sufficient to account for the negative geotaxis usually exhibited by this organism. The survival value of negative geotaxis for free-

swimming organisms in search of food may explain the characteristic shape of many protozoa.

Roberts, A. M. (1970*c*) Motion of *Paramecium* in static electric and magnetic fields. *J. Theor. Biol.* 27 : 97–106.

A quantitative model is constructed to explain the motion of *Paramecium* in an electric field. It is held that the membrane on the side of the cell closer to the cathode becomes depolarized and that this in turn renders the cilia less motile. The cilia on the anodal, hyperpolarized side are more active. Thus a moment derives from the asymmetry of propulsion, which leads to a spiral path of locomotion. Attention is also given to the effects of magnetic fields on the motility of the protozoan. It is unlikely that fields not exceeding 10^6 Oe can have a significant influence on the locomotion of *Paramecium*.

Rohlf, F. J. and Davenport, D. (1969) Simulation of simple models of animal behavior with a digital computer. *J. Theor. Biol.* 23 : 400–24.

The effects of kinetic, klinokinetic, orthokinetic, and tropotaxic behavior were simulated on a digital computer in order to study the interaction among these simple behaviors and the effect of sensory adaptation under completely controlled experimental conditions. The modeling was done without reference to any particular cell or organism type, and thus the results have complete generality. The conclusion of Ulyott (1936)[1] that klinokinetic behavior in the presence of sensory adaptation can cause a directional displacement of an organism along a gradient of stimulus was confirmed. In addition it was found that inverse klinokinetic and direct orthokinetic behavior (with sensory adaptation) result in a marked directional displacement up a gradient. However, direct klinokinetic behavior cancels out the effect of direct orthokinetic behavior.

1. Ulyott, P. (1936) *J. Exp. Biol.* 13 : 253.

Schreiner, K. E. (1971) The helix as propeller of microorganisms. *J. Biomechanics* 4 : 73–83.

A phenomenological approach is employed in determining the forces and moments on a rotating helix by the resistance of the surrounding liquid at low Reynolds numbers, and it is shown how they may be balanced by the forces and moments on a spherical head. Within the restrictions of the analysis it is found that if the head rotates it can have only one helical propeller. But if the number of flagella is $4n$, half of them rotating in each direction, the head will not rotate. The rate of energy transmission to the surrounding liquid is found to be lower in the latter case and to increase with the square of the ratio between helix and head radii in the former case. The efficiency of the propulsion system is found to be practically independent of the viscosity of the surrounding liquid.

Schreiner, K. E. (1972) *Sliding Filaments in Cilia and Flagella.* University of Oslo, Institute of Mathematics, Preprint Series in Applied Mathematics, no. 1., pp. 1–23.

The kinematics of idealized ciliary and flagellar motion are analyzed and it is shown how they correlate with the allowed discrete rates of motion in sliding filament mechanisms. The author proposes a model for the organelle motion according to which ATP is released in regions where active sliding and changes in orientation occur while regions of constant orientation are in rigor, in the same way as muscles under ATP deficiency. The progression rates of bends in planar organelle motion would then be determined by the rate at which crosslinks are transformed from an active state to a state of permanent attachment and rigor at the trailing end of the bend.

Silvester, N. R. and Holwill, M. E. J. (1972) An analysis of hypothetical flagellar waveforms. *J. Theor. Biol.* 35: 505–23.

This paper makes the interesting suggestion that flagellar waveforms may be of a meander type, that is, similar to meanders in a riverbed. One property of this form is that $\int_A^B (d\phi/ds)^2\, ds$ is a minimum for a constant path length extending between two points A and B, where ϕ is the angle between the tangent and the line connecting A and B, and s is the arc length. Previously, flagellar waveforms were described as sinusoidal or as circular arcs interspersed between straight regions.

The authors compare the three possible waveforms in two ways: (1) in geometry in order to find most critical parameters for experimentally identifying the waveform in real flagella, and (2) in propulsive force and efficiency that could be generated from the three waveforms.

Steinberg, M. S. (1963) Reconstruction of tissues by dissociated cells. *Science* 141: 401–08.

This paper reports a new, analytical approach to the problem of cell sorting out. Cells of various types are postulated to differ quantitatively in adhesivity rather than qualitatively. Sorting out is viewed as a result of the cells tending to assume a pattern of lowest potential energy, mediated by random motility and by the energies of cell–cell adhesion. Thus, cell sorting out is analogous to liquid-phase segregation. Depending on the quantitative relations between the works of adhesion binding the cell types present, sorting out can lead to complete separation of cell types, enveloping of some cell types by others, or complete intermixture. The model is extended to include situations where preaggregated cells are placed together *en bloc*, which may lead to spreading. The concepts here are impressively in line with experimental observations. This important analysis had been influential in later modeling of cellular motility, sorting out, and morphogenesis.

Steinberg, M. S. (1964) The problem of adhesive selectivity in cellular interactions. In *Cellular Membranes in Development* ed. M. Locke, pp. 321–66. New York: Academic Press.

This paper presents the author's differential adhesion hypothesis of cellular reaggregation specificity, which is described in Steinberg (1963)[1] (q.v.).This article also traces the roots of the theory to Holtfreter's earlier work. A shorter presentation of these ideas is given in Steinberg (1962).[2]

1. Steinberg, M. S. (1963) Tissue reconstruction by dissociated cells. *Science* 141: 401–08.
2. Steinberg, M. S. (1962) On the mechanism of tissue reconstruction by dissociated cells. III. Free energy relations and the reorganization of fused, heteronomic tissue fragments. *Proc. Nat. Acad. Sci. U.S.* 48: 1769–76.

Steinberg, M. S. and Roth, S. A. (1964) Phases in cell aggregation and tissue reconstruction. An approach to the kinetics of cell aggregation. *J. Exp. Zool.* 157: 327–38.

A theoretical assessment is made of the rate at which cells in suspension may collide with one another due to Brownian movement and to stirring if present. The effect on cell collision frequency of fibers that collect cells is also studied. Brownian motion cannot be responsible for significant aggregation at cell concentrations below 10^7 cells/ml.

Stirring can raise collision frequency at a rate approximately proportional to the third power of the cell diameter.

Stuhlman, O. (1948) A physical analysis of the opening and closing movements of the lobes of Venus' fly-trap. *Bull. Torrey Bot. Club* 75:22–44.

The kinetics and physical bases for closing and opening movements in the insectiv-orous plant traps are examined. The force of closure is also investigated to determine the site of application of the forces. It is concluded that the actively moving cells effect their behavior by means of regulating intracellular hydrostatic pressure and that the transduction of turgor pressure into mechanical movement is effected by mechanisms analogous to those operating in the curving of a Bourdon pressure gauge.

Subirana, J. A. (1970) Hydrodynamic model of amoeboid movement. *J. Theor. Biol.* 28: 111–20.

A hydrodynamic theory of the flow produced by active shear in amoebas is developed. Active shear is defined as the movement induced in a fluid at the interface with a resting solid. The monopodial movement of amoebas can be explained by this mechanism if it is assumed that active shear occurs between the inner surface of the gelled ectoplasmic tube and the outer surface of the fluid endoplasm. A similar mechanism may occur in other types of movement in different biological systems; their relative efficiencies are quantitatively compared in this paper.

Taylor, Sir G. (1951) Analysis of the swimming of microscopic organisms. *Proc. Roy. Soc. London A* 209: 447–61.

This is a pioneering analysis on the dynamics of movement by biological bodies under conditions of low Reynolds number. Assuming that movement is due to sinusoidal waves of a sheet, expressions are derived for velocity and propulsive thrust as functions of wave parameters, size, and medium viscosity. This model is valid for waves of low amplitude. It is also shown that two closely spaced, undulating bodies will be entrained to beat in unison, thus explaining such observed behavior in spermatozoa.

Thompson, D'Arcy W. (1942) *On Growth and Form.* 2nd ed. 2 vols. 1116 pp. Cambridge: Cambridge University Press.

This book contains many eloquent passages that bear strongly on cell movement and has been influential in the rise of mathematical biology. In general the book deals with statics rather than dynamics, and no single part may be construed as a discrete mathematical model of cell movement. The overriding importance of minimizing potential energies as a determinant of cell motility (see many articles in the present book) follows arguments similar to those presented in chapter 7 of *On Growth and Form*, entitled "The Forms of Tissues or Cell-Aggregates." Here the stable geometrical relationships in collections of cells are examined, often with reference to cells dividing or packed together.

Vasiliev, A. V. (1975) On a problem of sorting.[1] Probl. Peredachi Inform. (1971) no. 3 : 109–11.

The investigator considers the problem of sorting elements of two types packed into an n-dimensional lattice. At each discrete sorting step, one pair of differing neighboring elements is chosen at random. These elements exchange positions with a probability depending on the type of neighbors each has. This analysis explores the properties of the resulting Markov chain and finds the necessary and sufficient conditions for the existence of a final measure of Gibbsian type.

1. Paper 5 in this book.

Vasiliev, A. V. and Pyatetskii-Shapiro, I. I. (1971) Modeling of the process of sorting out cells on the computer. Sov. J. Develop. Biol. 2 : 286–91 (trans. of Ontogenez 2 : 356–62).

The process of cell sorting in mixed cell reaggregates is studied by computer simulation of a tessellation model similar to (but independent of) that of Goel et al. (1970).[2] The model is designed to see what parameters, in addition to adhesion energies, are needed to apply Steinberg's (1963) cell-sorting hypothesis[3] to real cells. Both adhesive energies and cell motility functions affect the degree of sorting out. Using strictly local rules and cell interaction, it is found that cells move a distance of only about 6 cell diameters regardless of the size of the tessellation.

2. Goel, N., et al. (1970) Self-sorting of isotropic cells. J. Theor. Biol. 28 : 423–68.
3. Steinberg, M. S. (1963) Reconstruction of tissues by dissociated cells. Science 141 : 401–08.

Vasiliev, A. B., Pyatetskii-Shapiro, I. I., and Radvogin, Y. B. (1972) Modeling of the processes of sorting out, invasion, and aggregation of cells.[4] Institute of Applied Mathematics of the Order of Lenin, Academy of Sciences of the USSR, Preprint 12 (1972).

The processes of cell sorting in heterotypic cell mixtures are examined by extensive computer simulation. The tessellation model employed is derived from Steinberg's (1963)[5] differential adhesion hypothesis and bears similarities to the similar models of Goel et al. (1970)[6] and Vasiliev and Pyatetskii-Shapiro (1971).[7] The present model incorporates parameters of variable distances of cell interaction, energy thresholds for cell movement, and bias of one cell's motility. The effects of varying these parameters and of altering cell proportions, adhesion energies, times of reaggregation, and sizes of reaggregates are explored. This paper substantiates the hypothesis that the characteristics of cell motility are of utmost importance in determining the effectiveness of sorting on the basis of adhesion energies.

4. Paper 4 in this volume.
5. Steinberg, M. S. (1963) Reconstruction of tissues by dissociated cells. Science 141 : 401–18.
6. Goel, N., et al. (1970) Self-sorting of isotropic cells. J. Theor. Biol. 28 : 423–68.
7. Vasiliev, A. V. and Pyatetskii-Shapiro, I. I. (1971) Modeling of the process of sorting out cells on the computer. Sov. J. Develop. Biol. 2 : 286–91.

Vul, E. B. and Pyatetskii-Shapiro, I. I. (1971). A model of inversion in *Volvox*. *Probl. Peredachi Inform.* 7, no. 4: 91–96.[1]

A model is constructed mimicking, in two dimensions, the inversion of *Volvox*. Cells operate by strictly local motility rules, and the algorithms are nearly homogeneous. This work represents an application of the more general analyses of line straightening and circularization studied previously.[2,3] Computer simulation of this model produced inversion figures similar to observed patterns in *Volvox*.

1. Paper 3 in this volume.
2. Leontovich, A. M., Pyatetskii-Shapiro, I. I., and Stavskaya, O. N. (1970) Certain mathematical problems related to morphogenesis. *Avtomatika i Telemekhanika* 4: 94–107.
3. Leontovich, A. M., Pyatetskii-Shapiro, I. I., and Stavskaya, O. N. (1971) The problem of circularization in mathematical modeling of morphogenesis. *Avtomatika i Telemekhanika* 2: 100–10.

Wang, C.-Y. and Jahn, T. L. (1972) A theory for the locomotion of spirochetes. *J. Theor. Biol.* 36: 53–60.

A hydrodynamic theory for the locomotion of spirochetes is presented. The theory is based on a possible arrangement of internal fibrils such that self-rotation about a local body axis is possible. This self-rotation is responsible for canceling the torque produced by the traveling helical waves of the body. The hydrodynamic theory follows the method used by Taylor (1952),[4] but with different boundary conditions. A relationship among the radius of body cross-section, amplitude, and wavelength of the helical wave is obtained for a spirochete traveling with no apparent slippage of water. The angular velocity of rotation about a local body axis is also determined. The theoretical results compare favorably with direct measurements of body geometry.

4. Taylor, G. (1952) The action of waving cylindrical tails in propelling microscopic organisms. *Proc. Roy. Soc. London A* 211: 225–39.

Weiss, L. (1964) Cellular locomotive pressure in relation to initial cell contacts. *J. Theor. Biol.* 6: 275–81.

In the amoeba *Chaos chaos*, the calculated electrostatic potential energy barrier to cell–cell contact is almost equaled by the energy available by the amoeba's forward thrust (locomotive pressure). Thus, in considering the mechanisms by which cells approach and touch one another, the investigator should take locomotive pressure into account.

Weston, J. A. and Roth, S. A. (1969) Contact inhibition: Behavioral manifestations of cellular adhesive properties *in vitro*. In *Cellular Recognition*, ed. R. T. Smith and R. A. Good, pp. 29–37, Appleton-Century-Crofts, New York.

Some aspects of the social behavior of cells in culture are hypothesized to depend on the relation between cell–cell and cell–substratum adhesive stability. The ratio of these two parameters is termed A. When A is high, cells will tend to cluster. When A is low, cells will tend to act independently. It is also hypothesized that at certain values of cell–substratum adhesive stability the cells become immobilized. Observations on the randomness of cultured cell distributions, as determined by the frequency of nuclear overlap, are consistent with this model.

Wilkie, D. (1954) The movements of spermatozoa of bracken (*Pteridium aquilinum*), *Exp. Cell. Res.* 6: 384–91.

This article contains a derivation of the rate at which fern spermatozoa should arrive at a trapping site by the process of random movements. It is used in this case for testing for nonrandom movement (chemotaxis) toward archegonial mucilage.

Williams, T. and Bjerknes, R. (1972) Stochastic model for abnormal clone spread through epithelial basal layer. *Nature* 236: 19–21.

This investigation examines the kinetics and pattern of clone expansion (to mimic tumor spread) in the basal layer of skin. It is assumed that spreading occurs by a dividing cell displacing a neighboring cell from the basal layer, resulting in that cell's loss by exfoliation. In the present model, the cells in question differ from normal by a decreased average cell cycle time (carcinogenic advantage). Using computer simulations of a hexagonal tessellation, the pattern of clonal spread was examined. One interesting finding is that the dimensionality of a clone's perimeter, where displacement of normal cells is occurring, is proportioned to $\sim N^{0.55}$ (N = number of cells in clone) and that this crinkliness of the perimeter is independent of N.

Yoneda, M. (1962) Force exerted by a single cilium of *Mytilus edulus*. II. Free motion. *J. Exp. Biol.* 39: 307–17.

The hydrodynamics of ciliary motion (whip-like rather than undulatory) is considered and an expression is derived for the force exerted by a cilium. Calculations are made to show the dependence of torque and power on the viscosity of the medium and on ciliary length.

Zigmond, S. H. and Hirsch, J. G. (1973) Leukocyte locomotion and chemotaxis. New methods for evaluation, and demonstration of a cell-derived chemotactic factor. *J. Exp. Med.* 137: 387–410.

Substances previously reported to elicit chemotaxis in leukocytes could be simply stimulating the rate of locomotion because most tests have not been able to distinguish these two effects on cell movement. The authors devise a method for distinguishing between increased rate of locomotion and orientation of locomotion. The method, based in part on a model of how fast cells would be expected to move in the absence of orientation but assuming increased velocity, shows that orienting factors do exist.